Software from

High Performance Systems, Inc.
The Systems Thinking Company™

An Introduction to Systems Thinking

STELLA® *Software*

High Performance Systems, Inc. • 45 Lyme Road, Suite 300 • Hanover, NH 03755
Phone: 603.643.9636 • Toll Free: 800.332.1202 • Fax: 603.643.9502
Technical Support: support@hps-inc.com • Pricing & Sales: sales@hps-inc.com
Workshop Info: workshops@hps-inc.com • To Order: orders@hps-inc.com
Visit us on the Web at: http://www.hps-inc.com
ISBN 0-9704921-1-1

Acknowledgements

This Guide was written by Barry Richmond. He received great support of various kinds from various people.

Nancy Maville, and Steve Peterson read and provided feedback on the Chapters. Steve also did a superb job of readying all of the models associated with the Guide, as well as creating the Index. Debra Gonzales formatted the text and Index, and also helped with rendering many of the Figures. Jeremy Wallis and Karim Chichakly programmed the features in the wonderful new version of the *STELLA* software upon which the Guide is based.

We hope you enjoy the fruits of our labor.

May, 2001

Contents

Part 1

The Language of Systems Thinking:

Operational, Closed-loop & Non-linear Thinking

We believe that constructing a good model using the *STELLA* software is very much analogous to writing a good composition, such as a short story, screenplay, or novel. And, because people have more familiarity with writing than they do with modeling, we've decided to rely pretty extensively on the analogy in hopes of accelerating your uptake of the modeling language, concepts, and process. Each of the remaining chapters in this Guide will draw upon the writing analogy.

As the title to this Part suggests, there is a parallel progression in the chapters that comprise it. One track is language. You'll begin, in Chapter 2, by learning the basic parts of speech in the stock/flow language. Chapter 3 will present the rules of grammar for constructing good sentences. In Chapter 4, you'll learn how to link sentences together. Chapters 5 and 6 will discuss how to compose first simple, then complex, paragraphs. Finally, Chapter 7 will illustrate how paragraphs can be put together to create a short story.

Paralleling the language track is the development of Systems Thinking skills. The chapters in this Part will focus on developing three key Systems Thinking skills: Operational, Closed-loop, and Non-linear Thinking.

The language and the thinking skills really are intertwined. You cannot write a good short story, or even compose a good sentence, unless you have a solid grasp of both the language and the associated thinking skills that enable you to apply it effectively.

Chapter 1

Systems Thinking and the *STELLA* Software:

Thinking, Communicating, Learning and Acting More Effectively in the New Millennium

I have been writing and re-writing this Guide for fifteen years. I always begin Chapter 1 by reeling off a litany of serious challenges facing humanity. And, you know what? The list has remained pretty much the same! There's homelessness and hunger, drug addiction and income distribution inequities, environmental threats and the scourge of AIDS. We've made precious little progress in addressing any of these issues over the last couple of decades! Indeed, you could make a strong case that, if anything, most (if not all) have gotten worse! And, some new challenges have arisen. Perhaps most disturbing among these is what appears to be (so far) largely an American phenomenon: kids killing kids (and teachers), at school.

So what's the problem? Why do we continue to make so little progress in addressing our many, very pressing social concerns?
My answer is that the way we *think*, *communicate*, and *learn* is outdated. As a result, the way we *act* creates problems. And then, we're ill-equipped to address them because of the way we've been taught to *think*, *communicate* and *learn*. This is a pretty sweeping indictment of some very fundamental human skills, all of which our school systems are charged with developing! However, it is the premise of this Chapter (and Systems Thinking) that it *is* possible to evolve our *thinking*, *communicating* and *learning* capacities. As we do, we will be able to make progress in addressing the compelling slate of issues that challenge our viability. But in order to achieve this evolution, we must overcome some formidable obstacles. Primary among these are the entrenched paradigms governing what and how students are taught. We *do* have the power to evolve these paradigms. It is now time to exercise this power!

I will begin by offering operational definitions of *thinking*, *communicating* and *learning*. Having them will enable me to shine light on precisely what skills must be evolved, how current paradigms are thwarting this evolution, and what Systems Thinking and the *STELLA* software can do to help. Finally, I'll overview what's to come in the remainder of the Guide. In the course of this Chapter, I will identify *eight* Systems Thinking skills. They are: 10,000 Meter,

System as Cause, Dynamic, Operational, Closed-loop, Non-linear, Scientific, and Empathic Thinking. Each will reappear, some receiving more attention than others, throughout the Guide. It is mastery of these skills that will enable you to make effective use of the *STELLA* software.

Providing Operational Definitions

The processes of thinking, communicating, and learning constitute an *interdependent system*, or at least have the potential for operating as such. They do not operate with much synergy within the current system of formal education. The first step toward realizing the potential synergies is to clearly visualize how each process works in relation to the other. I'll use the *STELLA* software to help with the visualization…

Thinking

Thinking…we all do it. But what is it? The dictionary says it's "…to have a thought; to reason, reflect on, or ponder." Does that clear it up for you? It didn't for me.

I will define thinking as consisting of two activities: *constructing* mental models, and then *simulating* them in order to draw conclusions and make decisions. We'll get to constructing and simulating in a moment. But first, what the heck is a *mental model*?

It's a "selective abstraction" of reality that you create and then carry around in your head. As big as some of our heads get, we still can't fit reality in there. Instead, we have *models* of various aspects of reality. We simulate these models in order to "make meaning" out of what we're experiencing, and also to help us arrive at decisions that inform our actions.

For example, you have to deal with your kid, or a sibling, or your parent. None of them are physically present inside your head. Instead, when dealing with them in a particular context, you select certain aspects of each that are germane to the context. In your mind's eye, you relate those aspects to each other using some form of cause-and-effect logic. Then, you *simulate* the interplay of these relationships under various "what if" scenarios to draw conclusions about a best course of action, or to understand something about what has occurred.

If you were seeking to understand why your daughter isn't doing well in arithmetic, you could probably safely ignore the color of her eyes when selecting aspects of reality to include in the mental model you are constructing. This aspect of reality is unlikely to help you in developing an understanding of the causes of her difficulties, or in drawing conclusions about what to do. But, in selecting a blouse for her birthday? Eye color probably ought to be in *that* mental model.

As the preceding example nicely illustrates, all models (mental and otherwise) are simplifications. They necessarily omit many aspects of the realities they represent. This leads to a very important statement that will be repeated several times throughout this Guide. The statement is a paraphrase of something W. Edwards Deming (the father of the "Quality movement") once uttered: "*All models are wrong, some models are useful.*" It's important to dredge this hallowed truth back up into consciousness from time to time to prevent yourself from becoming "too attached" to one of your mental models. Nevertheless, despite the fact that all models are wrong, *you have no choice but to use them*—no choice that is, if you are going to *think.* If you wish to employ non-rational means (like gut feel and intuition) in order to arrive at a conclusion or a decision, no mental model is needed. But, if you want to think...you can't do so without a mental model!

Figure 1-1 presents a *STELLA* map of the activities that comprise "thinking:" *constructing* (a mental model), and *simulating* in order to draw conclusions. As the Figure indicates, constructing is divided into two sub-activities: *selecting* and *representing*. The first sub-activity answers the question: *What should I include in my mental model?* The second sub-activity answers the question: *How should I represent what I include?* These are the two fundamental questions that must be answered in constructing *any* mental model. It is my conviction that the paradigms currently governing teaching in our schools restrict development of the whole set of skills needed to become effective in executing both the constructing and simulating activities. That is, our schools are thwarting development of thinking capacity—something no school board would approve, and we can ill afford!

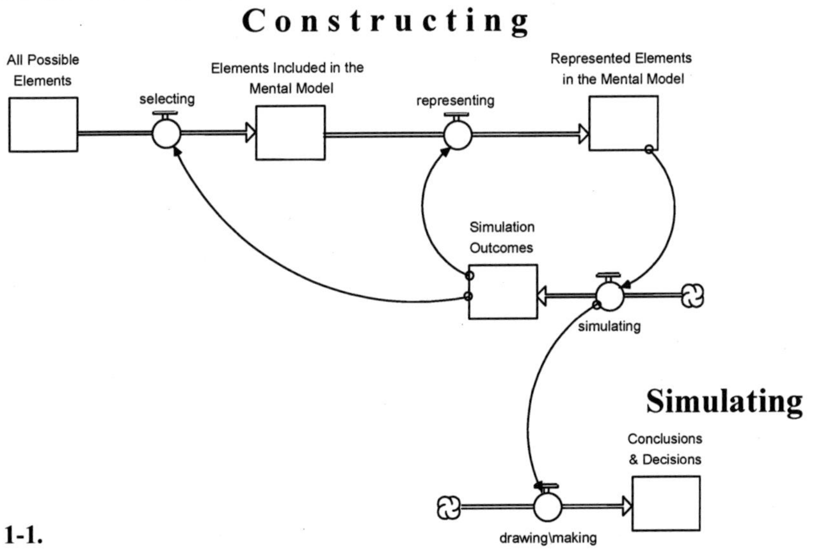

Figure 1-1.
A STELLA Picture of "Thinking."

The "wire" that runs from *Represented Elements in the Mental Model* to *simulating* is intended to suggest that *simulating* cannot proceed until a mental model is available—which is to say, the *selecting* and *representing* activities have been executed. Simulating yields conclusions that, among other things, help us to make decisions. But, as Figure 1-1 indicates, simulation outcomes play another important role in the thinking process. They provide feedback to the *selecting* and *representing* activities (note the "wires" running from *Simulation Outcomes* to the two activities). Simulation outcomes that make no sense, or are shown to have been erroneous, are a signal to go back to the drawing board. Have we left something out of our mental model that really should be in there, or included something that really doesn't belong? Have we misrepresented something we have included? This self-scrutiny of our mental models, inspired by simulation outcomes, is one of the important ways we all *learn*…but we're getting ahead in the story. Before we discuss *learning*, let's look at *communicating*.

Communicating

An operational picture of communicating is presented in Figure 1-2. The first thing to note is that the figure includes the elements that make up the *thinking* activity. The intention is to suggest that *communicating* is inextricably linked to *thinking*. Indeed, as the variable *Made Available for Scrutiny by Others* indicates, the outputs of the Thinking process provide the raw material for the Communicating process. Three sources of "raw material" are illustrated in the Figure: the mental model, the associated simulation outcomes, and the conclusions that have been drawn from simulating. By making these sources available, others then can "think" about them! Specifically, they can compare them to the corresponding information they possess. The comparison process, as you are about to see, drives a second type of *learning*!

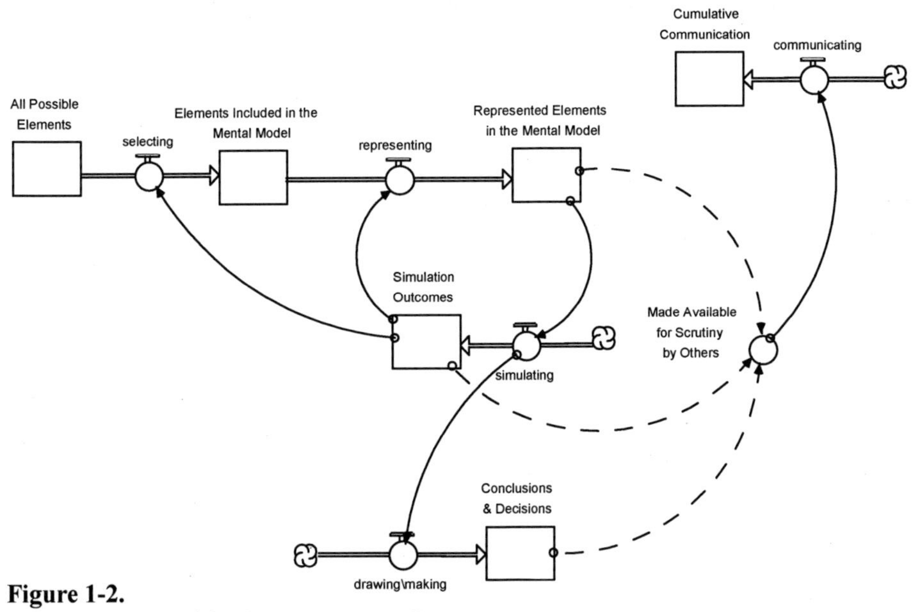

Figure 1-2.
A STELLA Map of the Communicating Process.

Learning

Learning is depicted in Figure 1-3. It's a pretty elaborate picture, and a good example of one that should be unfurled one chunk at a time using the *STELLA* software's storytelling feature, than sprung on you full-blown. If you *would* prefer to see the Figure 1-3 story "unfurled," open the model named "Learning" in the *Intro to Systems Thinking* folder, and the experience can be yours!

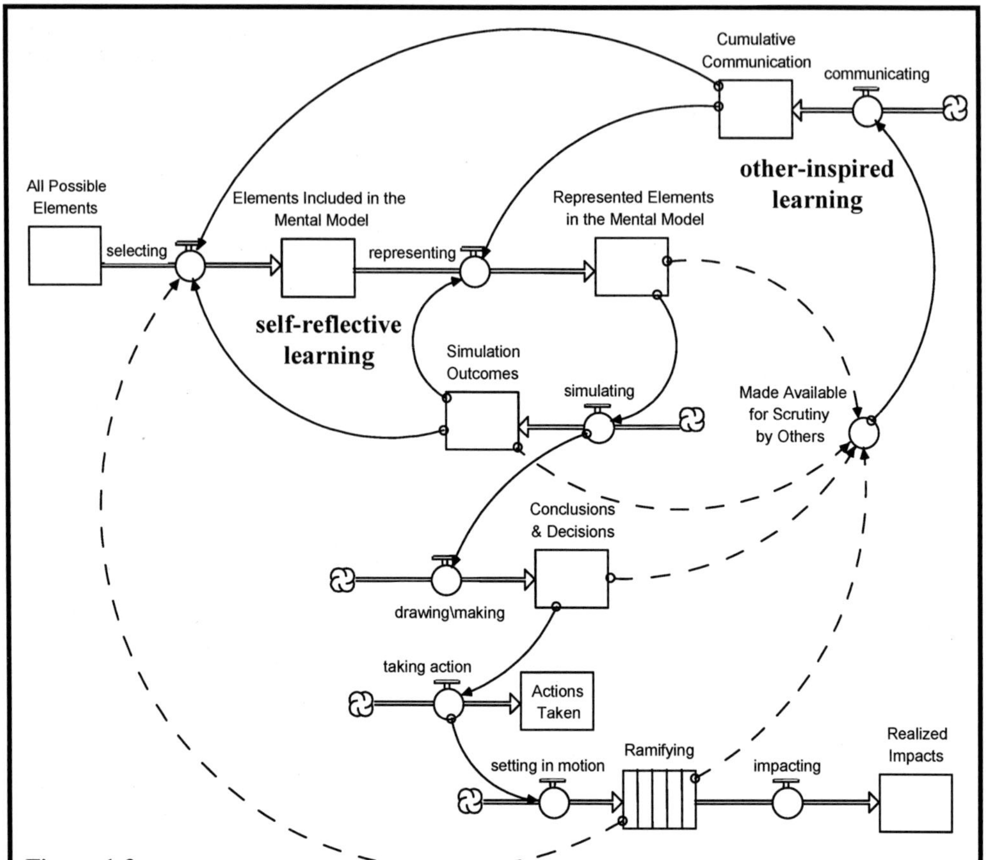

Figure 1-3.
A STELLA Map of the Learning Process.

The first type of learning was identified in the discussion of the *Thinking* process. Call it *self-reflective learning*. It comes about when simulation outcomes are used to drive a process in which a mental model's content, and/or representation of content, is changed. I've also just alluded to a second type of learning...one that's driven by the *Communicating* process. Call it *other-inspired* learning. As Figure 1-3 suggests, the raw material for this type of learning is: the mental model itself, the simulation outcomes associated with that model, and/or the conclusions drawn from simulating. How much learning occurs, depends upon both the quality of the feedback provided—where "quality" includes both content and "packaging"—as well as the willingness and ability to "hear" the feedback.

Figure 1-3 also adds a fourth source of raw material for learning: the impacts of one's actions. As the Figure suggests, often it is difficult to perceive the full impact because ramifying takes a long time, and spreads out over a great distance. To reflect this fact, the information for this type of learning is shown as radiating off the "conveyor" named *Ramifying*, rather than the stock called *Realized Impacts*. [NOTE: Conveyors are used to represent delays].

It's useful to spend a little time digesting Figure 1-3—which shows the thinking, communicating and learning *system*. An important thing to note about the Figure is that all roads ultimately lead back to learning—which is to say, *improving the quality of the mental model.* Learning occurs when either the content of the mental model changes (via the *selecting* flow), or the representation of the content changes (via the *representing* flow). By the way, to make the figure more readable, not all wires that run to the *representing* flow have been depicted.

There are two important take-aways from the Figure. First, the three processes—*thinking, communicating* and *learning*—form a self-reinforcing system. Building skills in any of the three processes helps build skills in *all* three processes! Second, unless a mental model changes, *learning does not occur*!

I will now use the preceding definitions of thinking, communicating and learning as a framework for examining how well the current system of formal education is preparing our youth for the issues they'll face as citizens in the new millennium. Wherever I indict the system, I'll also offer alternatives. The alternatives will emanate out of a framework called Systems Thinking, and make use of the *STELLA* software as an implementation tool. I'll begin with a blanket indictment, and then proceed using the thinking/communicating/learning framework to organize specific indictments.

The Blanket Indictment

If schools were mandated to pursue anything that looked remotely close to Figure 1-3, I wouldn't be writing this Chapter! Instead, students spend most of their time "assimilating content," or stated in a more noble-sounding way, "acquiring knowledge." And so, the primary learning activity in our schools is *memorizing*! It's flipping flash cards, or repeating silently to yourself over and over, the "parts of a cell are…," the "three causes of World War II are…," the "planets in order away from the sun are…" Students cram facts, terms, names, and dates in there, and then spit them back out in the appropriate place on a content-dump exam. This despite the fact that students perceive much of the content to have little perceived relevance to their lives, and that a good chunk of the content will be obsolete before students graduate.

Notice something about the process of "acquiring knowledge." It bears no resemblance to the process depicted in Figure 1-3. In acquiring knowledge, no mental model is constructed. No decisions are made about what to include, or how to represent what's included. No mental simulating occurs. Acquiring knowledge also doesn't require, or benefit from, communicating. Quite the contrary, the knowledge acquisition process is solitary, and *non-thinking* in nature. And then, the coup de gras...Will *content* really equip our young people for effectively addressing the issues they'll face in the new millennium?

It's important to recognize that although I am indicting the content-focus of our education system, I am not indicting the teachers who execute that focus (at least not all of them)! Pre-college teachers, especially, are hamstrung by rigid State (and in some cases, Federal) mandates with respect to material to be taught, pedagogic approach, and even sequencing. My indictment is primarily aimed at the folks who are issuing these mandates! I'm indicting those who have established measurement systems that employ a content-recall standard for assessing mastery, and who confuse "knowing" with "understanding" and "intelligence." To you, I wish only to say (loudly): *Wake Up!*

That said, let's get on with some specific indictments, and with suggestions for doing something to improve the situation.

Thinking: *Constructing a Mental Model*

Whether the mental model being constructed is of an ecosystem, a chemical reaction, a family, or a society, three fundamental questions must always be answered in constructing it. They are: (1) What elements should be included in the model—or, the flip side—what elements should be left out? (2) How should the elements you decide to include be represented? (3) How should the relationships between the elements be represented?

What to Include?

Deciding what to include in a mental model, in turn, breaks into two questions. *How broadly do you cast your net?* This is a "horizontal" question. And, *how deeply do you drill?* This is a "vertical" question. Developing good answers to these two questions requires skill. And, like any skill, this one must first be informed by "good practice" principles, and then honed through repeated practice. Let's see how development of the "what to include?" skills fares in the current education system.

The first thing to note is that little time remains for developing such skills because so much time is allocated to stuffing content—which as noted, is an activity that does not require "what to include/how to represent" choices. Nevertheless, the formal education system *does* leave its stamp on *selection* skills. And, it's not a particularly useful one!

One of the implicit assumptions in the prevailing educational paradigm is that what's knowable should be *segmented*. The rationale appears to be that it will enable content to be assimilated most efficiently. The resulting student learning strategy might be called: "Divide & Conquer." Those who are best at executing this strategy reveal their expertise at mid-term and final time, effecting a serial, single-content focus—e.g., putting assimilated history content aside, in order that it not interfere with imbibing biology content. Over time, students figure out which content areas they're "best at," and then concentrate on these. The result is that students become *content specialists*. At the same time populations of math-phobics, literature-phobics, language-phobics, and science-phobics are created. Students come to see the world as divided into "content bins," some of which they "like," others of which, they avoid.

Content specialists tend to cast their nets narrowly (over the domains they "know"). And, they also tend to focus their gaze deeply—they've stored lots of detail about their "comfort" arena(s). Their mental models thus tend to be narrow and deep. They contain a lot…about a little. Meanwhile, students' skills in seeing *horizontal* connections never really develop. Instead, vertical detail dominates big picture.

The problem with this approach to developing student thinking capacity is that *all* of the challenges I ticked off at the start of the Chapter—homelessness, income distribution inequity, global warming, AIDS, kids killing kids, etc.—are *social* in nature! They arise out of the interaction of human beings with each other, with the environment, with an economy. They are problems of *interdependency*! They are horizontal problems! That's because the horizontal boundaries of social systems, in effect, go on forever. Make a change within a particular organization, for example, and the ripple effects quickly overflow the boundaries of the organization. Each employee interacts with a raft of people outside the organization who, in turn, interact with others, and so on. So, in the social domain, being able to think *horizontally* is essential! Nets must be cast broadly, before drilling very deep into detail. Yet, to the extent students' selection skills are being developed at all, they are being biased in exactly the *opposite* direction…toward bin-centricity.

Systems Thinking offers three thinking skills that can help students to become more effective in answering the "what to include" question. They are: "10,000 Meter," "Systems as Cause," and "Dynamic" Thinking.

10,000 Meter Thinking

The first thinking skill, 10,000 Meter Thinking, was inspired by the view one gets on a clear sunny day when looking down from the seat of a jet airliner. You see horizontal expanse, but little vertical detail. You gain a "big picture," but relinquish the opportunity to make fine discriminations.

System as Cause Thinking

The second Systems Thinking skill, "System as Cause" Thinking, also works to counter the vertical bias toward including too much detail in the representations contained in mental models. "System as Cause" thinking is really just a spin on Occam's razor (i.e., the *simplest* explanation for a phenomenon is the *best* explanation). It holds that mental models should contain only those elements whose interaction is capable of *self-generating* the phenomenon of interest. It should not contain any so-called "external forces." A simple illustration should help to clarify the skill that's involved.

Imagine you are holding slinky as shown in Figure 1-4a. Then, as shown in Figure 1-4b, you remove the hand that was supporting the device from below. The slinky oscillates as illustrated in Figure 1-4c. The question is: *What is the cause of the oscillation?* Another way to ask the question: *What content would you need to include in your mental model in order to explain the oscillation?*

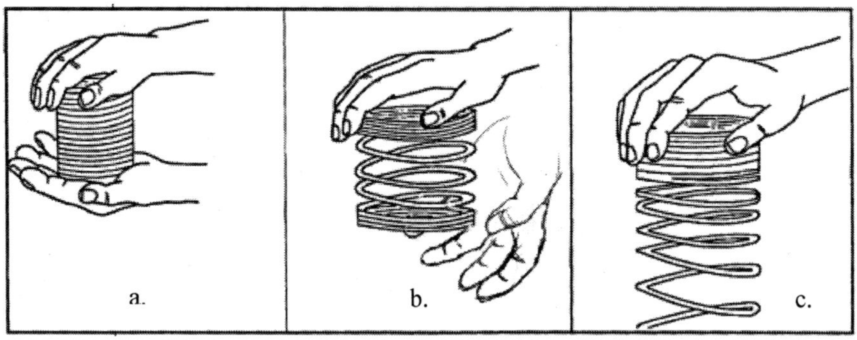

Figure 1-4.
A Slinky Does Its Thing.

The two, most common causes cited are: gravity, and removal of the hand. The "System as Cause" answer to the question is: the slinky! To better appreciate the merits of this answer, imagine that you performed the exact same experiment with, say, a cup. The outcome you'd get makes it easier to appreciate the perspective that the oscillatory behavior is latent within the structure of the slinky itself. In the presence of gravity, when an external stimulus (i.e., removing the supporting hand) is applied, the dynamics latent within the structure are "called forth." It's not that gravity and removal of the hand are irrelevant. However, they wouldn't appear as part of the "causal content" of a mental model that was seeking to explain why a slinky oscillates.

Dynamic Thinking

The third of the so-called "filtering skills" (Systems Thinking skills that help to "filter" out the non-essential elements of reality when constructing a mental model) is called "Dynamic Thinking." This skill provides the same "distancing from the detail" that 10,000 Meter Thinking provides, except that it applies to the behavioral—rather than the structural—dimension.

Just as perspectives get caught-up in the minutiae of structure, they also get trapped in "events" or "points," at the expense of seeing patterns. In history, students memorize dates on which critical battles were fought, great people were born, declarations were made, and so forth. Yet in front and behind each such "date" is a pattern that reflects continuous build-ups or depletions of various kinds. For example, the US declared its independence from England on July 4, 1776. But prior to that specific date, tensions built continuously between the two parties to the ensuing conflict. In economics, the focus is on equilibrium *points*, as opposed to the trajectories that are traced as variables move between the points.

Dynamic Thinking encourages one to "push back" from the events and points to see the pattern of which they are a part. The implication is that mental models will be capable of dealing with a dynamic, rather than only a static, view of reality.

Figure 1-5 should help make clearer the difference between the "Divide & Conquer"-inspired viewpoint and the Systems Thinking-inspired perspective in terms of the resulting content of a mental model. The Figure makes the contrast between mental models constructed using the alternative perspectives look pretty stark. That's an accurate picture. Yet there is nothing to prevent models forged using *both* perspectives from co-existing within a single individual. Nothing, that is, but finding room for developing the three associated Systems Thinking skills (10,000 Meter, System as Cause, and Dynamic Thinking) in a curriculum already overstocked with mandated discipline-focused "knowledge acquisition" requirements. To be sure, there have always been (and will always be) efforts made to develop horizontal thinking skills, usually in the form of cross-disciplinary offerings. But such efforts are scattered, and rely heavily on the "extra-curricular" commitment and enthusiasm of particular individuals. And, they grow increasingly rare as grade levels ascend, being all but non-existent at the post-secondary level.

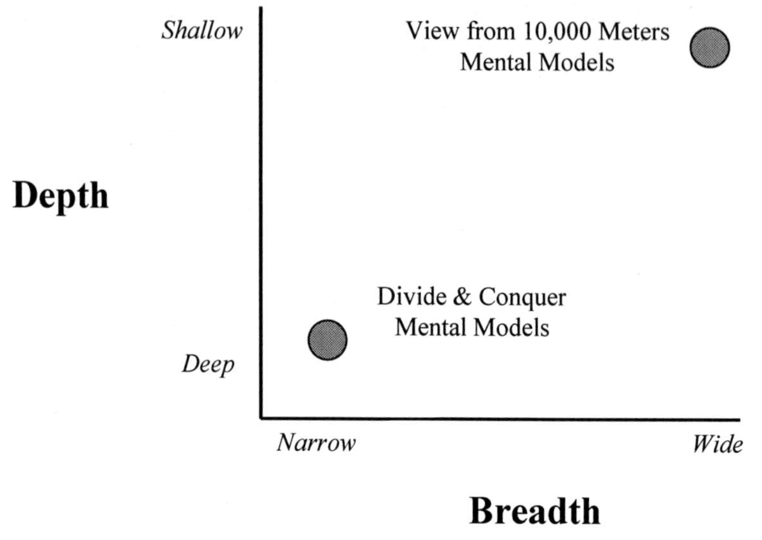

Figure 1-5.
The Content of Divide & Conquer-inspired Versus Systems Thinking Mental Models.

Until the average citizen can feel comfortable embracing mental models with horizontally-extended/vertically-restricted boundaries, we should not expect *any* significant progress in addressing the pressing issues we face in the social domain. And until the measurement rubrics on which our education system relies are altered to permit more focus on developing horizontal thinking skills, we will continue to produce citizens with predilections for constructing narrow/deep mental models. The choice is ours. Let's *demand* the change!

How to Represent What You Include

Once the issue of *what* to include in a mental model has been addressed, the next question that arises is how to represent what has been included. A major limit to development of students' skills in the representation arena is created by the fact that each discipline has its own unique set of terms, concepts, and in some cases, symbols or icons for representing their content. Students work to internalize each content-specific vocabulary, but each such effort contributes to what in effect becomes a *content-specific* skill.

Systems Thinking carries with it an icon-based lexicon called the language of "stocks and flows." This language constitutes a kind of *Esperanto*, a lingua franca that facilitates *cross-disciplinary* thinking and hence implementation of a "horizontal" perspective. Mental models encoded using stocks and flows, *whatever the content*, recognize a fundamental distinction among the elements that populate them. That distinction is between things that accumulate (called "stocks") and things that flow (called "flows"). Stocks represent conditions within a system—i.e., how things are. Flows represent the activities that cause conditions to change. Some examples of accumulations are: water in a cloud, body weight, and anger. The associated flows are: evaporating/precipitating, gaining/losing, and building/venting. Figure 1-6 should help you to develop a clearer picture of the distinction between a stock and a flow.

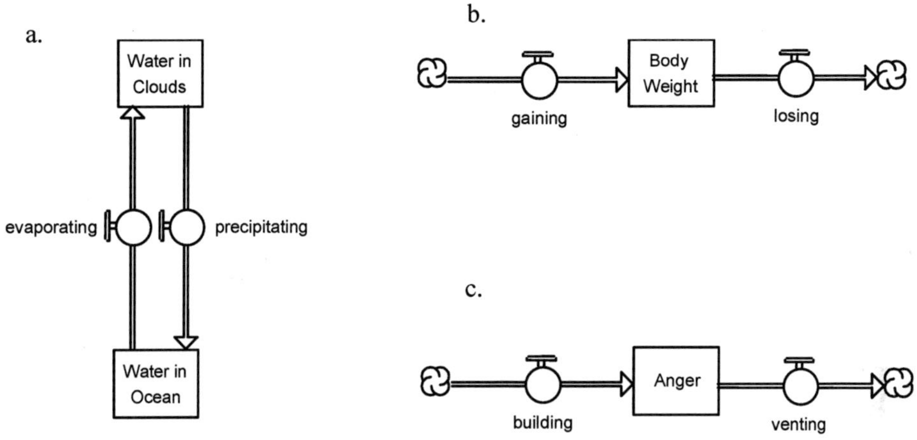

Figure 1-6.
Some Stocks & Flows.

To gain a quick idea of why the distinction matters, consider the illustration in Figure 1-6b. Suppose a person whose weight has been increasing, decides to take some action to address the situation. First, they successfully eliminate *all* junk food snacks from their diet, and do not eat more at regular meals to compensate for doing so. Second, they implement a rigorous aerobic exercise program—to which they religiously adhere. This means the person will have lowered the volume of the *gaining* flow (i.e., reduced caloric intake) and increased the volume of the *losing* flow (increased caloric expenditure).

So what happens to this person's body weight?

Did your answer include the possibility that it would still be increasing? It should have! Look at Figure 1-6b. The reason the person may still be gaining weight is because decreasing the rate of *gaining* (the inflow), and increasing rate of *losing* (the outflow), will only cause *Body Weight* (the stock) to decrease if *gaining* actually drops below *losing*. Until this occurs, the person will continue to gain weight—albeit at a slower rate! Take a moment to make sure you understand this reasoning before you proceed.

When the distinction between stocks and flows goes unrecognized—in this example, and in any other situation in which mental simulations must infer a dynamic pattern of behavior—there is a significant risk that erroneous conclusions will be drawn. In this case, for example, if the inflow and outflow volumes do not cross after some reasonable period of time, the person might well conclude that the two initiatives they implemented were ineffective and should be abandoned. Clearly that is not the case. And, just as often, the other type of erroneous conclusion is drawn: "We're doing the right thing, just not enough of it!" Redoubling the effort, in such cases, then simply adds fuel to the fire.

In addition to helping increase the reliability of mental simulations, using stocks and flows in representing the content of a mental model has another very important benefit. The benefit derives from the fact that the concepts of accumulation and flow are *content-independent.* Therefore, in whatever specific content arena they are used, the use contributes to building the *general* content-representation skill! Figure 1-7 seeks to capture this idea via the links that run from each of four content-specific representing activities to the building of a general content-representation skill.

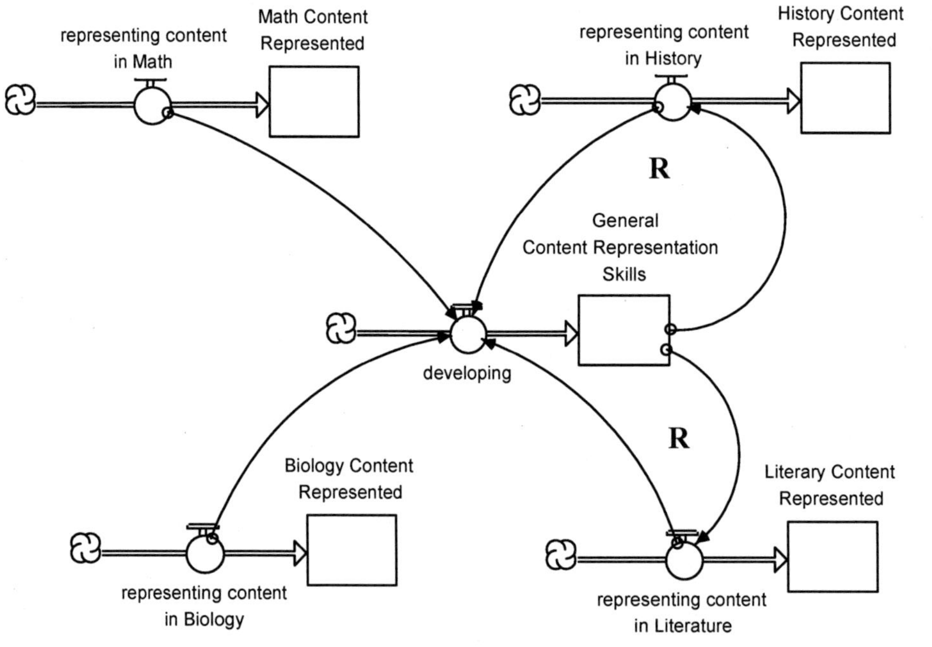

Figure 1-7.
Developing General Content Representation Skills by Representing Specific Content.

There's a second important idea illustrated in Figure 1-7. Note the two "R's." They stand for the word "Reinforcing"—which is the type of *feedback loop* they designate. The loops work like this…

As *general* content representation skills build, they facilitate each specific content-representing activity—though, to keep the picture simple, the link to only two of the specific arenas is illustrated. Then, as students engage in *specific* content-representation activities, because they are using a content-transcendent language to do so, they develop *general* content representation skills—a *virtuous learning cycle*! The cycle creates synergy because *all* content arenas benefit from activities that go on in *any one* of them! Now, instead of one content arena *interfering* with learning in another, each helps to accelerate learning in each of the others.

Operational Thinking

To be able to "speak/write" effectively in the language of stocks and flows requires that students build a fourth Systems Thinking skill, a *very* important one: *Operational Thinking*. Much of Chapters 2-7 are taken up with developing this skill, so I'll not say any more about it here. Teaching the language of stocks and flows, and the associated Operational Thinking skills, at an early point in the formal education process (e.g., fourth, fifth, sixth grade) would be a huge step toward enabling students to develop a better set of *representing* skills. It would, at the same time, leverage development of students' horizontal thinking skills. And the good news is that at the lower grade levels, there still remains sufficient flexibility in many curricula to permit taking this step. *Carpe diem*!

How to Represent the Relationships Between Elements that You Include

The final question we must answer in constructing a mental model is how to represent the *relationships* between the elements we decide to include. In answering this question, we must necessarily make some assumptions about the general nature of these relationships. Among the most sacred of all the covenants that bind members of a society together is the implicit agreement about how such relationships work. In Western cultures, the implicit agreement is that reality works via a structure of serial cause-and-effect relationships. Thus-and-such happens, which leads this-and-such to occur, and so forth. Not all cultures "buy" serial cause-and-effect (some subscribe to perspectives such as "synchronicity" and "God's hand"). But Western culture does.

I have no beef with serial cause-and-effect. It's a useful viewpoint. However, when I look more closely at the assumptions that characterize the particular brand of it to which Western culture subscribes, I discover that these assumptions seriously restrict learning! Let's see how…

The name I use for the Western brand of serial cause-and-effect is "Laundry List Thinking" (another name would be "Critical Success Factors Thinking"). Laundry List Thinking is defined by a set of four "meta" assumptions that are used to structure cause-and-effect relationships. I use the term "meta" because these assumptions are *content-transcendent.* That is, we use them to structure cause-and-effect relationships whether the content is Literature, Chemistry, or Psychology, and also when we construct mental models to address personal or business issues. Because we all subscribe to these "meta" assumptions, and have had them inculcated from the "get go," we are essentially unaware that we even use them! They have become so obviously true they're not even recognized as assumptions any more. Instead, they seem more like attributes of reality.

But, as you're about to see, the "meta" assumptions associated with Laundry List Thinking are likely to lead to structuring relationships in our mental models in ways that will cause us to draw erroneous conclusions when we simulate these models. I will identify the four "meta" assumptions associated with Laundry List Thinking, and then offer a Systems Thinking alternative that addresses the shortcomings of each. Here's a question that I'll use to surface all four assumptions…

What causes students to succeed academically? Please take a moment and actually answer the question.

Before I proceed with harvesting the question, I want to provide some evidence to suggest the Laundry List framework is in very widespread use both in academic and non-academic circles.

On the non-academic side, "recipe" books continue to be the rage. One of the first, and most popular, of these is Stephen Covey's *The Seven Habits of Highly Effective People*. The habits he identifies are nothing more (nor less) than a laundry list! And, for those of you familiar with the "critical success factors" framework, it, too, is just another name for a laundry list. In the academic arena, numerous theories in both the physical and social sciences have been spawned by Laundry List Thinking. For example, one very popular statistical technique known as "regression analysis," is a direct descendent of the framework. The "Universal Soil Loss" equation, a time-tried standard in the geological/earth sciences, provides a good illustration of a regression analysis-based, Laundry List theory. The equation explains erosion (A, the *dependent* variable) as a "function of" a list of "factors" RKLSCP (the *independent* variables):

A=RKLSCP

A soil loss /unit of area
R rainfall
K soil erodibility
L slope length
S slope gradient
C crop management
P erosion control practice

Okay, so now that I've provided some evidence that Laundry List Thinking is quite widespread, you shouldn't feel bad if you (like most people) produced a laundry list in response to the "*What causes students to succeed academically?*" question.

If you did produce such a list, it probably included some of the variables shown on the left-hand side of Figure 1-8. The Figure belies *four* "meta" assumptions about cause-and-effect relationships implicit in the Laundry List framework. Let's unmask them!

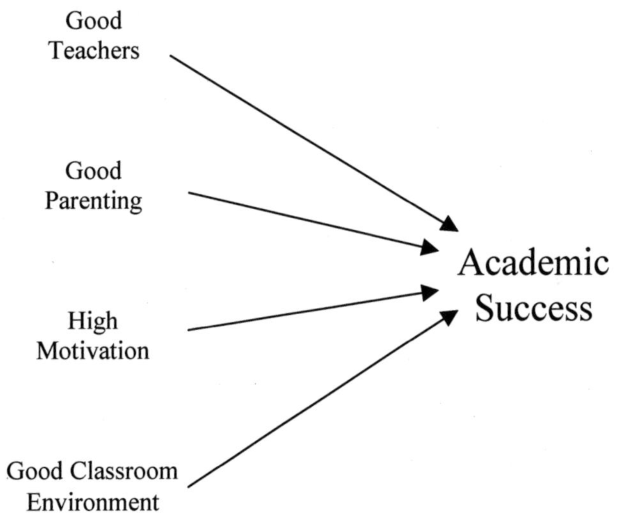

Figure 1-8.
A Laundry List Thinking Mental Model.

The first "meta" assumption is that the causal "factors" (four are shown in Figure 1-8) each operate *independently* on "the effect" ("academic success" in the illustration). If we were to "read the story" told by the view depicted in the Figure, we'd hear: "Good Teachers cause Academic Success; Good Parenting cause…" Each factor, or independent variable, is assumed to exert its impact *independently* on Academic Success, the *dependent* variable.

To determine how much sense this "independent factors" view really makes, please consult *your* experience…

Isn't it really a "partnership" between teachers and parents (good open lines of *reciprocal* communication, trust, etc.) that enables both parties to contribute effectively to supporting a student's quest for academic success? And don't good teachers really help to create both high student motivation and a good classroom environment? Isn't it the case that highly motivated students and a good classroom environment make teaching more exciting and enjoyable, and as a result cause teachers to do a better job? I could continue. But I suspect I've said enough to make the point. The four factors shown in Figure 1-8 aren't even close to operating *independently* of each other! They operate as a tightly intertwined set of *interdependent* relationships. They form a *web of reciprocal causality*! The picture that emerges looks much more like Figure 1-9, than Figure 1-8!

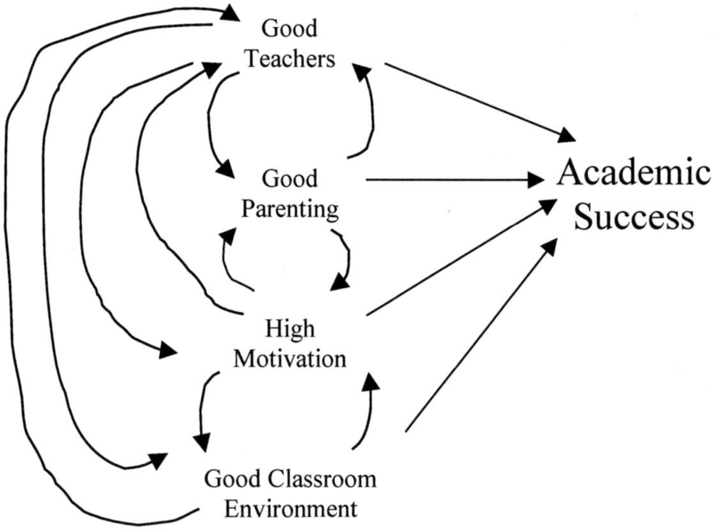

Figure 1-9.
From Independent Factors to Interdependent Relationships.

So, there goes the first "meta" assumption associated with Laundry List Thinking (i.e., that the causal "factors" operate *independently*). Now let's watch the second Laundry List "meta" assumption bite the dust! The second assumption is that *causality runs one-way*. Look back at Figure 1-8. Notice that the arrows all point from cause to effect; all run from left to right. Now steal another glance at Figure 1-9. Notice anything different?

That's right, the arrows linking the "causes" now run *both* ways! Cause-and-effect comes in *loops*! As Figure 1-10 shows, once circular cause-and-effect enters the picture, the so-called "effect" variable also loses its "dependent" status. It, too, now "causes"—which is to say that academic success stimulates student motivation and a good classroom environment, just as much as they drive it. Academic success also causes teachers to perform better—it's easier to teach students who are doing well—just as much as good teachers create academic success. And so forth. "Academic Success" is just as much a *cause* of any of the four "factors" as they are a cause of it! And so, independent and dependent variables become chickens and eggs. Everybody becomes a co-conspirator in a causal web of interrelationships.

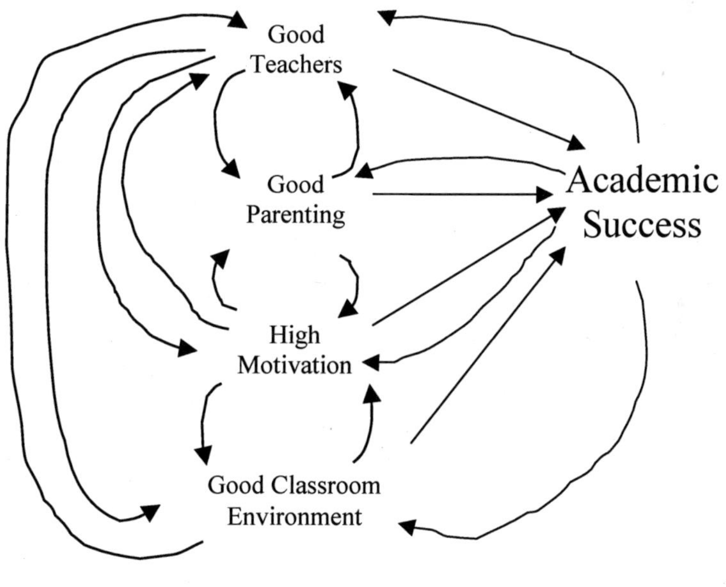

Figure 1-10.
Effect is also Cause.

Closed-loop Thinking

The shift from the Laundry List—causality runs one-way—view, to System Thinking's two-way, or closed-loop, view is a big deal! The former is *static* in nature, while the latter offers an "ongoing process," or *dynamic*, view. Viewing reality as made up of a web of *closed loops* (called *feedback loops*), and being able to structure relationships between elements in mental models to reflect this, is the fifth of the Systems Thinking skills. It's called *Closed-loop Thinking.* Mastering this skill will enable students to conduct more reliable mental simulations. Initiatives directed at addressing pressing social issues will not be seen as "one-time fixes," but rather as "exciting" a web of loops that will continue to spin long after the initiative is activated. Developing closed-loop thinking skills, will enable students to better anticipate unintended consequences and short-run/long-run tradeoffs. These skills also are invaluable in helping to identify high-leverage intervention points. The bottom line is an increase in the likelihood that the next generation's initiatives will be more effective than those launched by our "straight-line causality"-inspired generation.

The third and fourth "meta" assumptions implicit in Laundry List Thinking are easy to spot once the notion of feedback loops enters the picture. The causal impacts in Laundry Lists are implicitly assumed to be "linear," and to unfold "instantaneously" (which is to say, without any significant delay). Let's examine these two remaining Laundry List "meta" assumptions...

Non-linear Thinking

The assumption of "linearity" means that each causal factor impacts the "effect" by a fixed, proportional magnitude. In terms of the Universal Soil Loss equation, for example, someone might collect data for a particular ecosystem and then statistically estimate that, say, an 8% increase in rainfall (R) results in a 4% increase in soil loss per unit of area (A). We could then form the following equation to express the relationship: $A = 0.5R$. You probably immediately recognized it as your old friend...the equation of a straight line (i.e., $Y = mX + b$). In a linear equation, a given change in the "X" variable results in a fixed corresponding change in the "Y" variable. The variable expressing the amount of the corresponding change is "m," the *slope* of the straight line relating the two variables. Let's contrast the "linear" view of the relationship between rainfall and soil loss, with a "non-linear" view as illustrated in Figure 1-11.

As the wire running from *raining* to *eroding away* shows, erosion is "driven by" rainfall. The equation for *eroding away* is *raining* (an amount of water per time) times *soil lost per unit of water*. Notice the "~" on the face of the variable named *soil lost per unit of water*. It designates the variable as what's called a "graphical function." (I will discuss the graphical function in more detail in Chapter 6). The function is drawn as a graph on the right side of Figure 1-11. The graphical relationship indicates that the amount of soil that flows away with each unit of water is *not* constant! Instead, it depends upon the amount of *Vegetative Cover* that's present at the time. In particular, as the amount of *Vegetative Cover* increases, the quantity of *soil lost per unit of water* decreases—an *inverse* relationship (vegetation sinks roots into the soil that help to hold soil particles together, and in so doing, reduces erosion).

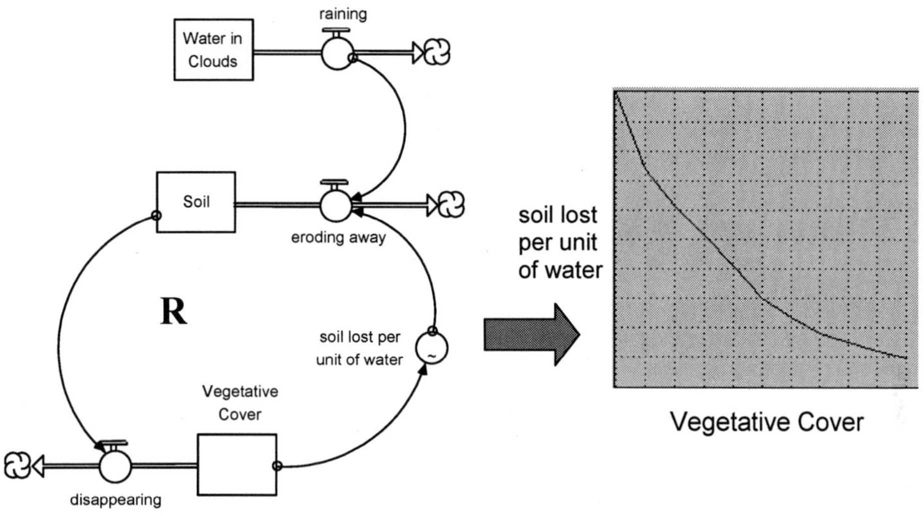

Figure 1-11.
A "Non-linear" Look at Soil Erosion.

The assumption being made here is that there is *not* a "linear" relationship between rainfall and the amount of soil being carried away by water. Instead, the strength of the relationship will *change* as the magnitude of a third variable, *Vegetative Cover*, changes. And, the plot thickens…As the wire running from *Soil* to *disappearing* (the outflow from *Vegetative Cover*) indicates, the rate at which vegetative cover disappears depends on how much soil is in place. The less soil in place, the more rapidly vegetative cover disappears; the more rapidly vegetative cover disappears, the less of it there is; the less vegetative cover, the more rapidly soil will be lost. A vicious cycle, or *Reinforcing feedback loop* (thus the "R").

Feedback loops, as they interact with waxing and waning strength, create *non-linear* behavior patterns—patterns that frequently arise in both natural and social systems. Such patterns cannot arise out of simulations of mental models whose relationships are *linear*.

Developing *Non-linear Thinking* skills (the sixth of the Systems Thinking skills) will enable students to construct mental models that *are* capable of generating such patterns. This, in turn, will enable students to better anticipate the impacts of their actions, as well as those of the initiatives that will be implemented to address the pressing social and environmental concerns they will face upon graduation.

The fourth implicit "meta" assumption associated with Laundry List Thinking is that impacts are felt "instantaneously." For example, when we look at the factors impacting academic success, the implicit assumption is that each exerts its influence "right now." Take "Good Classroom Environment." The idea here is that a good classroom environment—i.e., physical factors like space, light, good equipment, etc.—will encourage students to achieve high levels of academic success. Boost the quality of the physical environment...boost academic success. Sounds reasonable, but when you draw a more operational picture, the cause-and-effect is not quite so straight-forward. Take a look at Figure 1-12.

Instead of words and arrows—Good Classroom Environment → Academic Success—to show causality, Figure 1-12 depicts the associated causal relationships *operationally*. In particular, the Figure includes the potentially significant *delay* between initiating improvements to a classroom environment and the "arrival" of those improvements. The vehicle for capturing the delay, as you've already seen (in Figure 1-3), is called a "Conveyor." In this illustration, suppose the delay had to do with, say, the delivery, and subsequent bringing on line, of a mobile computer lab for the classroom. Such delays have been known to stretch out for months. In the mean time, it's possible that student and teacher morale might suffer. This, in turn, could stimulate an *outflow* from the *Level of Academic Success* before the arrival of the new lab has a chance to stimulate the associated inflow!

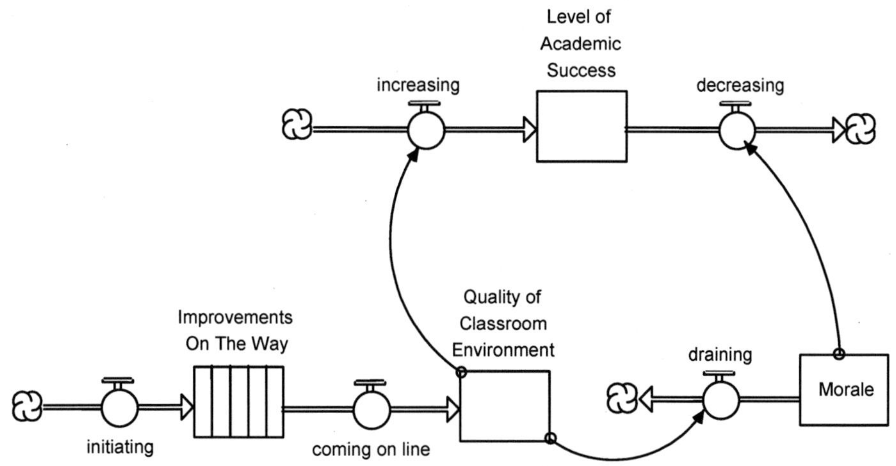

Figure 1-12.
A "Non-instantaneous" View.

Delays are an important component of how reality works. Leaving them out when structuring relationships in mental models undermines the reliability of simulation outcomes produced by those models. Building the *Operational Thinking* skills that enable students to know when and how to include delays should be a vital part of any curriculum concerned with development of effective thinking capacities.

A Brief Recap

Okay, it's been a long journey to this point. Let's briefly recap before resuming. I asserted at the outset that our education system was limiting the development of our students' *thinking*, *communicating* and *learning* capacities. I have focused thus far primarily on thinking capacities. I have argued that the education system is restricting both the *selecting* and *representing* activities (the two sub-processes that make up *constructing* a mental model). Where restrictions have been identified, I have offered a Systems Thinking skill that can be developed to overcome it. Six Systems Thinking skills have been identified thus far: 10,000 Meter, System as Cause, Dynamic, Operational, Closed-loop and Non-linear Thinking. By developing these skills, students will be better equipped for constructing mental models that are more congruent with reality. This, by itself, will result in more reliable mental simulations and drawing better conclusions. But we can do even more!

We're now ready to examine the second component of thinking, *simulating*. Let's see what's being done to limit development of students' capabilities in this arena, and what we might do to help remedy the situation.

Thinking: *Simulating a Mental Model*

The first component of thinking is *constructing* mental models. The second component is *simulating* these models. Throughout the discussion thus far, I've been assuming that all simulating is being performed *mentally*. This is a good assumption because the vast majority is performed mentally. How good do you think you are at mental simulation? Here's a test for you…

Read the passage that follows and then perform the requested mental simulation …

A firm managing a certain forestland is charged with maintaining a stable stock of mature trees, while doing some harvesting of trees each year for sale. Each year for the last 50 years or so, the firm has harvested a constant number of mature trees. In order to maintain the stock of mature trees at the specified target level, the firm follows a policy of re-planting a seedling for each mature tree it harvests in a given year. In this magically ideal forest preserve, no animals eat seedlings, and *every* seedling that is planted not only survives, but grows to maturity in *exactly* six years. Because the preserve has been operating in this manner for more than 50 years, it is in "steady-state." This means that an equal (and constant) number of trees is being harvested each year, an equal number of seedlings is being planted each year, and that same number of trees is also maturing each year. The stock of mature trees has therefore remained at a constant magnitude for 50 years.

Now, suppose that this year the firm decides to step up the harvesting of mature trees to a new, higher rate, and to then hold it constant at this rate for the foreseeable future.

Mental simulation challenge: *If the firm continues with its current re-planting policy (i.e., re-plant one seedling for each mature tree that it harvests), and ideal conditions for seedlings continue to prevail in the preserve, what pattern, over time, will be traced by the magnitude of Mature Trees following the step-increase in the harvesting rate?* Sketch your guess on the axis provided in Figure 1-13.

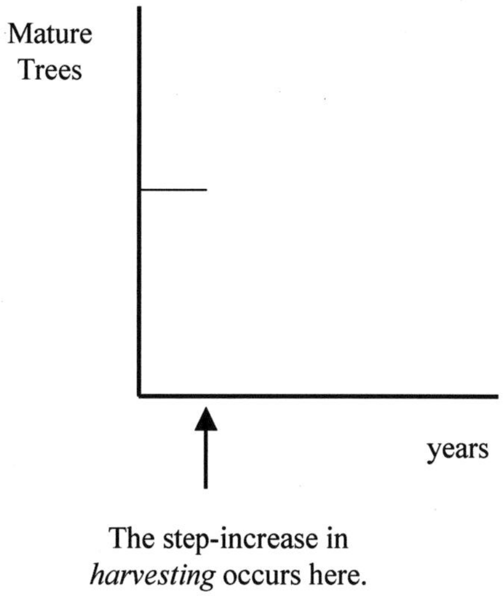

Figure 1-13.
Your Guess?

If you are like 90% of the people to whom we've put a question like this, you sketched an incorrect pattern. If you'd like to check your intuition, open the model named "Trees" in your *Intro to Systems Thinking* folder and run it.

The fact that 90% of the people who take this test guess incorrectly is significant. The percentage holds cross-culturally, and independently of gender, education level, and any other attribute we've looked at. This means the result is saying something about human beings *in general*! It's saying that, as a species, we're not very good at constructing a mental model from a written description, and/or mentally simulating that model once it is constructed. It's worth noting that the system we asked you to model and simulate is very simple! It's a whole lot simpler, for example, than the one spitting up issues like kids killing kids, drug addiction, and global warming. And we're simulating this latter system in our heads in order to create policy initiatives for addressing these issues! Scary? You bet!

If you refer back to Figure 1-3, you'll be reminded that *simulating* is a key part of the self-reflective learning loop. Reflecting on the simulation outcomes we generate is an important stimulator of change in our mental models. But what if those outcomes are bogus? What if we are not correctly tracing through the dynamics that are implied by the assumptions in our mental models? That's right...The *Self-reflective* learning loop will break down. In addition, because simulation outcomes are one of the raw materials being made available for scrutiny by others in the communicating process, a key component of the *Other-inspired* loop will break down, as well. So, it's very important that our simulation results be reliable in order that the associated learning channel can be effective.

Detailing the reasons for our shortcomings (as a species) in the simulation sphere is beyond the scope of this Chapter. However, part of the issue here is certainly biological. Our brains simply have not yet evolved to the point where we can reliably juggle the interplay of lots of variables in our heads. There is, however, growing evidence to suggest that people can hone this capacity. But in the current education system, there is very little attention being paid to this vital skill.

Systems Thinking can offer a couple of things that can help in this arena. The first is the language of stocks and flows. Because the language is both visual, and operational, it facilitates mental simulation. As an illustration, look at Figure 1-14. It's a *STELLA* map developed from the tree-harvesting story. Let's use it to facilitate a mental simulation.

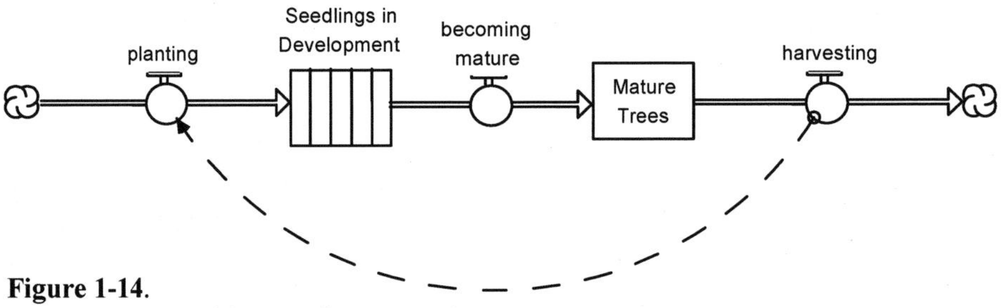

Figure 1-14.
A STELLA Map of the Tree-harvesting Story.

As described in the written passage, the system begins in steady-state. This state is easy to visualize using the map. It means that the two stocks are constant, because the three flows are equal (and also constant). The *harvesting* flow then steps-up to a new, higher level and remains there. Given this pattern for the outflow from *Mature Trees*, the map "tells you" that the pattern over time traced by the stock will be completely determined by what happens to the *becoming mature* flow. Do you "hear" this?

If the *becoming mature* flow steps-up at the same time as the *harvesting* flow, the *Mature Trees* stock will remain unchanged; i.e., inflow and outflow will remain equal. Hence, the magnitude of the stock will not change. But does the *becoming mature* flow step up at the same time as the *harvesting* flow?

No! For six years after the step-increase in *harvesting* occurs, the *becoming mature* flow will remain equal to the pre-step *harvesting* rate. That's because there is six year's worth of seedlings that are "in development," and the number of seedlings in each year's cohort is equal to the value of the pre-step *harvesting* rate. So, six years *after* the step increase in harvesting occurs, the *becoming mature* flow will finally step-up to equal the new, higher volume of *harvesting.* At this point, the system will be back in steady-state. However, because the *becoming mature* flow volume was less than the *harvesting* volume for six years, the stock of *Mature Trees* will have declined for six years. And because *becoming mature* was less than *harvesting* by a constant amount, the decline will be *linear*. The *Mature Trees* stock will now rest at a *permanently lower* level than existed prior to the step-increase in *harvesting*.

Scientific Thinking

STELLA maps really do facilitate mental simulation! But the other nice thing about them is that they are readily convertible into models that can be simulated by a computer. And if you follow "good practice" in doing your *STELLA* simulations, they will serve as an excellent "sanity-check" on your mental simulation. Think of the software as a fitness center for strengthening mental simulation "muscles." In order to take full advantage of the exercise facility, it's important to acquire the habit of making explicit a guess about what dynamics a particular model will generate *before* actually using *STELLA* to generate them. Experience has shown that it is far too easy to "back rationalize" that you "really knew" the model was going to produce that pattern. It's also important to put your models into steady-state (at least initially), and to test them using "idealized test inputs" (like STEP and PULSE functions). The collection of rigorous simulation practices are called *Scientific Thinking*, the seventh of the Systems Thinking skills.

Thinking, In Summary

Currently, in the formal education system, very little attention is paid to developing simulation skills. This means that a very important set of feedback loops for improving the quality of mental models is essentially being ignored. The *STELLA* software is a readily available tool that can play an important role in helping to develop these skills.

Communicating

The next process in the Thinking/Communicating/Learning system is *Communicating*. The kind of communicating I'm talking about here is not restricted to what one usually learns in an English composition class. The communicating I'm talking about must become a vital part of *every* class! It's the feedback students provide after scrutinizing each other's mental models and associated simulation outcomes (refer to Figure 1-3).

The current formal education system provides few opportunities for students to share their mental models and associated simulation outcomes. Well-run discussion classes do this (and that's why students like these classes so much!). Students sometimes are asked to critique each other's writing, or oral presentations, but most often this feedback is grammatical or stylistic in nature.

Empathic Thinking

The capacity for both giving and receiving feedback on mental models is vital to develop if we want to get better at bootstrapping each other's learning! Many skills are involved in boosting this capacity, including listening, articulating, and, in particular, *empathizing* capabilities. Wanting to empathize increases efforts to both listen and articulate clearly. Being able to empathize is a skill that can be developed—and is in some ways, the ultimate Systems Thinking skill because it leads to extending the boundary of true caring beyond self (a skill almost everyone could use more of). By continually stretching the horizontal perspective, Systems Thinking works covertly to chip away at the narrow self-boundaries that keep people from more freely empathizing.

But even with heightened empathic skills, we need a language that permits effective across-boundary conversations in order for communication to get very far. And this is where the issue of a content-focused curriculum resurfaces as a limiting factor. Even if time were made available in the curriculum for providing student-to-student feedback on mental models, and empathy were present in sufficient quantity, disciplinary segmentation would undermine the communication process. Each discipline has its own vocabulary, and in some cases, even its own set of symbols. This makes it difficult for many students to master all of the dialects (not to mention the associated content!) well enough to feel confident in, and comfortable with, sharing their reflections. The stock/flow *Esperanto* associated with Systems Thinking can play an important role in raising students' level of both comfort and confidence in moving more freely across disciplinary boundaries.

Figure 1-15 illustrates this notion...

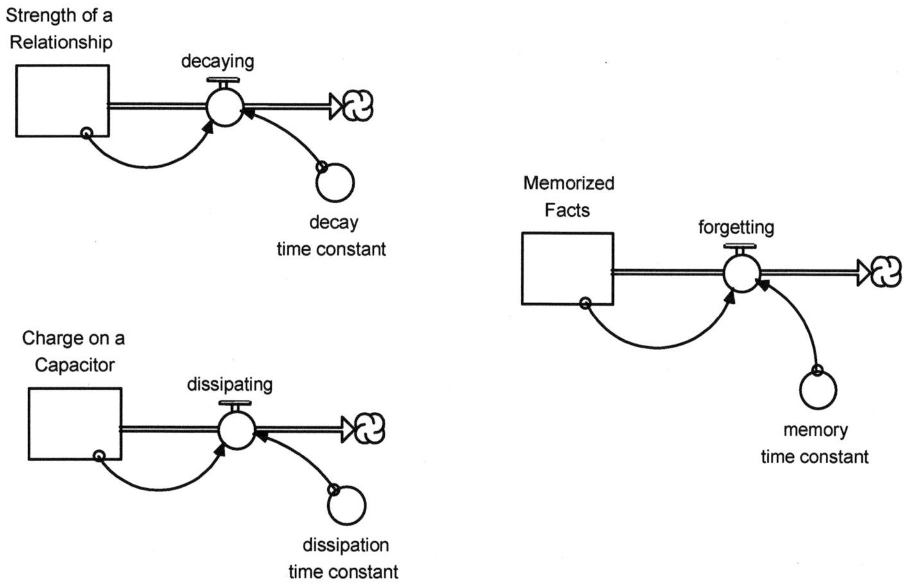

Figure 1-15.
The Generic Structure of a Dissipation Process.

The Figure shows the accumulation of strength in a personal relationship, the accumulation of electrostatic charge on a capacitor, and the accumulation of facts in human memory. Each is represented by the *same* symbol. As stocks, each performs an analogous function—albeit in quite different contexts—which is to "report" the status of a condition. In addition, as illustrated in the Figure, the "logic" by which one or more of the associated flows operate is *generic*. This is, at the very least, a comforting discovery in a world generally perceived to be growing more complex and unfathomable on a daily basis, and in a curriculum rife with detail-dense, dialect-specific content bins. But it also holds the wonderful potential for creating cross-curricular learning synergies. What's being learned in physics could actually *accelerate* (rather than impede) learning in literature or psychology (and vice versa)! And, by building their capacity for seeing "generic structures," students will be simultaneously boosting their capacity for making "horizontal" connections in the real world.

Teaching the stock/flow *Esperanto*, and the *Operational and Empathic Thinking* skills needed to "speak/write it" effectively, can go a long way toward improving the student communication capacities needed to realize the synergies latent within a multi-discipline curriculum.

Chapters 2-9 of this Guide should provide the nucleus of what's required to deliver this instruction.

Learning

Learning is both literally and figuratively the "bottom line" of the thinking/communicating/learning triumvirate. Because the three processes are intimately intertwined, all of the Systems Thinking and *STELLA*-based suggestions that have been made for improving the *thinking* and *communicating* processes would also work to improve the *learning* process. There is one more suggestion I would like to make that focuses more exclusively on the learning process itself. I'll enlist the *STELLA* software to paint what I hope will be a clear picture of the suggestion. The picture appears as Figure 1-16.

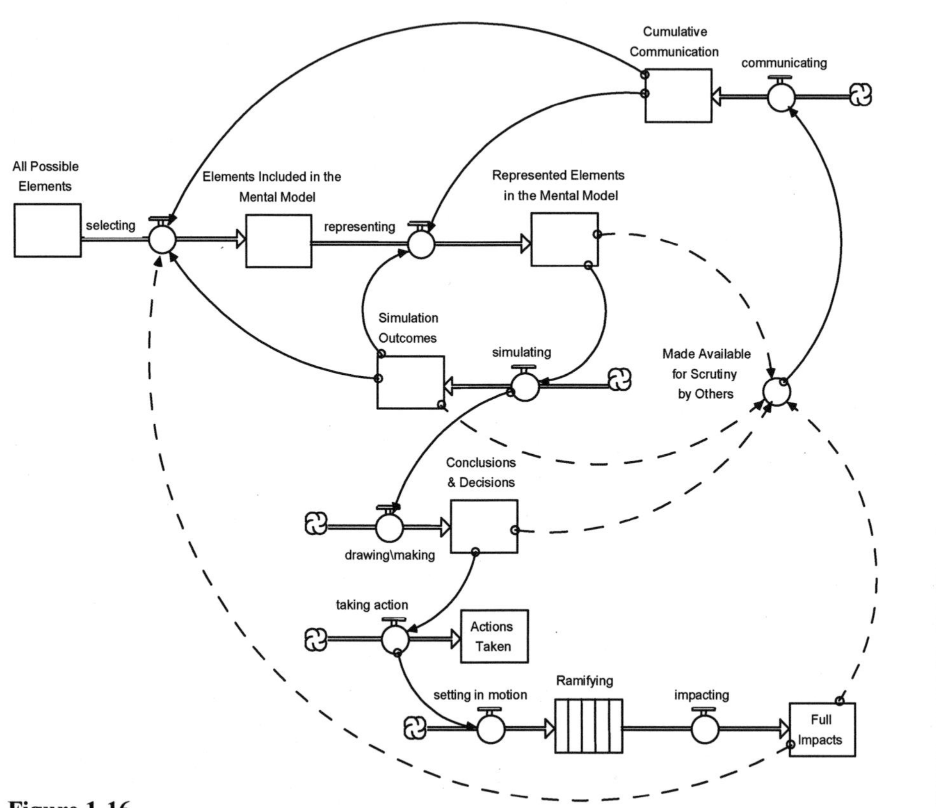

Figure 1-16.
Capturing the Full Impacts of Actions.

In processing the Figure, you may wish to take a look back at Figure 1-3. It differs from Figure 1-16 in only one way. The two learning-generation links, which emanated from *Ramifying* in Figure 1-3, now come off *Full Impacts*. This implies that somehow we've been able to "close the learning loop" on the full ramification of actions that have been taken, rather than capturing only the partial impacts (because those impacts were still ramifying). How might we be able to achieve this?

The answer I'd like to propose falls under the rubric of what's known as "organizational learning." This is a term, tossed about with abandon, which has been deeply enshrouded in fog since it was first coined. To borrow a phrase...Organizations don't learn, *people* do! I use the term "organizational learning" to refer to *learning that is captured, and then somehow stored, outside the bodies of the individuals who create and make use of it.* As such, when individuals disappear, their contribution to the collective understanding does not go with them. And, when new people arrive, they are able to quickly come up to the current collective level of understanding because that understanding is housed in some extra-corporal reservoir.

The vehicle I would propose for creating this "extra corporal" reservoir—call it an "organizational learning infrastructure"—is a set of *STELLA* models. The infrastructure would work as follows...Each model would be used to predict what will occur (not in a numerically precise way, but in a qualitative sense) in whatever context it is serving. A process would be in place to monitor actual outcomes versus model-generated predictions. When discrepancies between the two arise, the assumptions in the model would be scrutinized, discussed, and then adjusted accordingly. Over time, the model would continuously improve as a representation of the reality about which learning is being accumulated.

It would be great to implement this sort of "extra corporal" learning process in a classroom over a school year, perhaps even extending it to multiple years—and thereby giving students some sense of learning continuity as they progress through grade levels. Having developed experience with such a process while in school may inspire some students to continue the much-needed practice of seeking to harvest the learning from "full impacts" in their professional and public service careers.

In Summary

The challenges today's students will face when they leave school are formidable, and growing more so every day. The education system has not evolved its curriculum, methods, and tools so as to better equip students for addressing these issues. The system continues to be driven by a "content acquisition" standard that features *memorization* as its primary "learning" activity. The key to evolving our education system lies in tapping the potential synergies that exist in the mutually-reinforcing processes of *thinking*, *communicating* and *learning*. Systems Thinking and the *STELLA* software can bring a lot to this party!

This Chapter identified eight Systems Thinking skills that leverage all three processes. Each skill can be readily implemented into today's school systems. The primary barrier to doing so is the view that the mission of an education system is to fill students' heads with knowledge. This view leads to sharp disciplinary segmentation and to student performance rubrics based on discipline-specific knowledge recall. Changing viewpoints—especially when they are supported by a measurement system and an ocean of teaching material—is an extremely challenging endeavor. But the implications of not doing so are untenable. The time is now.

What's to Come

The remainder of the Guide relies on an extended analogy. Learning to use the *STELLA* software to render mental models is treated as analogous to learning to write an expository composition, such as a short story or screenplay. The Guide is divided into two parts.

Part 1 is entitled **The Language of Systems Thinking:** *Operational, Closed-loop, and Non-linear Thinking.* The six chapters in this Part form a parallel progression of language/grammar and the associated thinking skills needed to apply that language and grammar effectively. You'll build up from parts of speech to short story themes, and in the process begin to internalize the first three of the eight Systems Thinking skills.

Part 2 of the Guide is entitled **The Writing Process**: *10,000 Meter, System as Cause, Dynamic, Scientific and Empathic Thinking.* In the three chapters in this Part, you'll learn good "writing" practices, walk through an illustration of these practices, and finally be given some general "writing" guidelines.

As you've probably concluded if you've endured to this point, this isn't your typical "User's Manual." That's because learning how to make effective use of the *STELLA* software really has little to do with the mechanics of the software itself. The software's user interface is simple enough to master just by "playing around" for a few hours. The real issue with the *STELLA* software is internalizing the associated Systems Thinking skills, as well as the language and method. This is conceptual, not mechanical, work! The Guide is concerned with helping you to make a shift of mind, and to internalize a new language. If you need technical assistance in learning to use the software, there are excellent *Online Help Files* and self-study tutorials that accompany your software. For conceptual help, visit the HPS website (www.hps-inc.com) for articles and references to Systems Thinking resources.

Congratulations on your purchase of the *STELLA* software, and good luck in your efforts to apply it. The benefits you'll reap from learning Systems Thinking will re-pay many times over the investment you will make!

Chapter 2

Nouns & Verbs

Operational Thinking

Most languages recognize the fundamental distinction between nouns and verbs. The *STELLA* language is no different. Nouns represent things and states of being; verbs depict actions or activities. As we'll see in the next chapter, it takes at least one noun and one verb to constitute a grammatically correct "sentence" in the *STELLA* language, just as it does in other languages. So we're on very familiar ground with this language. The big difference is that the *STELLA* language icons are *operational* in nature. This means that when you tell a story using them, you can see it not only with your mind's eye, but also with your *real* eyes! And everyone else can see it with *their* real eyes, too. *Operational* means "telling it like it really is." And when you do, ambiguities and chances for miscommunication are greatly reduced. You wouldn't want to compose sonnets for your loved one using the *STELLA* language. But if you're trying to make explicit your mental model of how something actually works, you just can't beat it!

Nouns

Nouns represent things, and states of being. The "things" can be physical in nature, such as: Population, Water, Cash, and Pollution. They also can be non-physical in nature, such as: Quality, Anger, Hunger, Thirst, Self-esteem, Commitment, and Trust. Non-physical things are often "states of being." A theme that emerges early on in Systems Thinking is the *full-citizen status* that is accorded to non-physical variables. The *STELLA* software is just as applicable in Literature, Philosophy, Sociology, Psychology, and Anthropology, as it is in Physics, Biology, Chemistry, or Engineering.

Nouns in the *STELLA* language are represented by rectangles. The rectangle was chosen for a good reason. Rectangles look like bathtubs viewed from the side. And bathtubs turn out to be a good, physically intuitive metaphor for what all nouns represent: i.e., *accumulation.* That's right, accumulation! Cancer cells pile up in a tumor. Cash builds up in bank accounts. Anger builds up all over your body—adrenalin levels in your bloodstream, blood pressure, tension in your muscles. Love swells over the course of a relationship. So, when you think about nouns in the *STELLA* language, and you see rectangles, think of them as bathtubs that fill and drain. The difference is that these "tubs" will only rarely contain water.

Nouns, in the *STELLA* language, are called "stocks." The convention in naming stocks in the *STELLA* language is to designate them with first-letter capitalization. As you'll see, this will help in visually distinguishing them from flows—which typically are scripted in all lower-case letters.

There are four varieties of stocks: *reservoirs, conveyors, queues, and ovens*. The *Help Files* do an exquisite job of documenting the functioning of each. Here, our task will be to help you distinguish the four types, and to determine when it is most appropriate to use each.

By far, the most frequently used type of stock is the *reservoir*. You can use a reservoir to perform essentially all of the functions of any of the other types of stock. A distant second in frequency of use is the *conveyor*. And way back there, almost in total obscurity, are the *queue* and *oven* (which is why these two are not part of the basic software offering, the *Basic* version, and are included only in the *Research* version). The lineup of stocks appears in Figure 2-1.

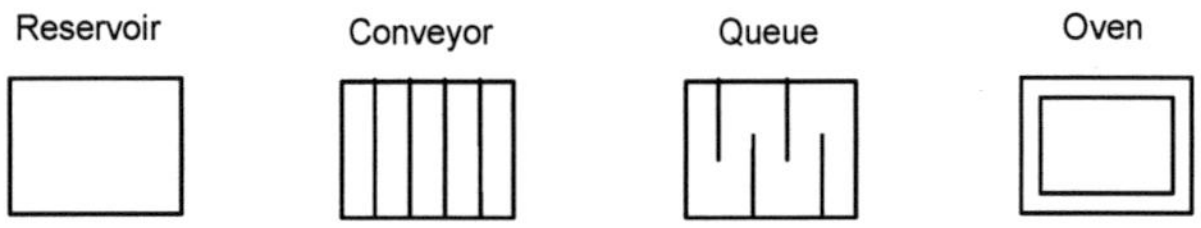

Figure 2-1.
The Four Types of Stock.

The Reservoir

The *reservoir* operates most like a real bathtub. Individual entities flow into a reservoir, and then become indistinguishable—just as individual water molecules flowing into a bathtub become indistinguishable (i.e., you can't tell which molecule arrived first, which tenth, and which arrived last). Instead, the molecules blend together; all arrival time discipline and size-uniqueness are lost. You just have a certain number of liters of water in the tub. The same is true when you use a reservoir to represent, say, Population or Cash. You can't distinguish Jamal from Janice in a reservoir labeled Population. You just have a total number of people. And the $100 bills are indistinguishable from the $1,000 bills in a reservoir named Cash. You just have a total amount of money. You can't tell which bill came in when, nor can you distinguish bills of different denominations. That's what reservoirs do. They blur distinctions between the individual entities that flow into and out of them. Instead, they collect whatever *total* volume of stuff flows in, and give up whatever *total* volume flows out. At any point in time, they house the net of what has flowed in, minus what has flowed out.

The Conveyor

Think of conveyors as like those "moving sidewalks" at O'Hare or Heathrow airports. Or, conjure up an escalator at your favorite mall or department store. You step on either, you stand and ride for some distance, you get off—unless you're one of those Type A's who has to walk at full stride (while being transported) so as to at least *double* your ground speed. That's how conveyors work. Whatever quantity arrives at the "first slat" gets on. It occupies the "first slat" on the conveyor. Nothing else can occupy that slat. The quantity "rides" until the conveyor deposits it "at the other end." The "trip" will take a certain amount of time to complete (known as the "transit time"). Conveyors are great for representing "pipeline delays" and all varieties of "aging chains."

Unlike reservoirs, conveyors *do* maintain arrival integrity and, sometimes, also batch size. If one $100 bill arrives at the "first slat" at time 3, and one $500 bill arrives at time 5, you'd be able to distinguish the bills while they're on the conveyor, and the $500 bill will "get off" two time units after the $100 bill—assuming the transit time of the conveyor remains constant (an assumption that can be relaxed—see the *Online Help Files* for details). Batch size is *not* retained in situations where, say, two $100 bills arrive at time 3 (you'd then simply have a total quantity of $200 "riding along").

The "danger" in relying too heavily on conveyors, a danger that heightens when employing queues and ovens, is loss of the 10,000 Meter viewpoint—a key viewpoint needed to do effective Systems Thinking. When you begin distinguishing between individual trucks, and worrying about whether that particular one (the red one over there) was delivered at 9:15 or 9:17, you have descended into the weeds and will no longer be able to see "the big picture." You're looking for specific answers, not general insights. You've traded your compass for a detailed street map. And you're also pushing the boundaries of what the *STELLA* software is best suited for doing. As a general rule, try to use reservoirs. If they really won't do the job, go with a conveyor. If you find yourself "going with a lot of conveyors," call us, we'll schedule you a "10,000 Meter" experience.

The Queue & The Oven

Frankly, we included these "mutants" in the software because the very technical end of the population using the software asked for them. These elements are pretty important for doing what's called "discrete event" simulations. Don't worry if this term is foreign to you. Suffice it to say that the *STELLA* software emanates out of a fundamentally "continuous" viewpoint on reality—again, we are talking the "10,000 meter" view. Queues and ovens serve the "discrete" worldview. Including them in the software represents our attempt to do what physicists have been trying to do for 150 years—resolve the wave/particle duality issue! We figured, "No problem guys, here's the answer you've been looking for!"

Queues

This said, for certain applications, queues and ovens can be useful. So I'll briefly describe them here. A queue is a "line" like you often see waiting to check in at an airline ticket counter, or in front of our offices every morning waiting to purchase the *STELLA* software. Queues develop when things arrive at a rate that exceeds the capacity to "process" them. Think of cars stacking up at the tollbooths on the George Washington Bridge, waiting to enter New York City. Or, even closer to my own heart, cars amassing at one of the multiple entrances to what New Englanders affectionately refer to as "a rotary" (and I refer to as "the circle of death"). Ah, civility at its best!

Queues retain both arrival integrity and batch size. In the *STELLA* software, queues enforce niceness. No "cutting in line" or "saving a place for a friend" is allowed. There's also no "leaving" once you're in line. When a volume of stuff "arrives," if it can't "get in/get on," it sits in a queue (in a unique spot) until it can. Stuff that arrives later "gets in line" behind the stuff that's already there. And it stays there! Again, you can visit the *Online Help Files* for more information on Queues—and remember that this stock type is available only in the *Research* version of the software.

Ovens

If conveyors are escalators, ovens are elevators. People arrive at an elevator, and if the doors happen to be open, they enter and then ride. In the more likely event that the doors are closed...people queue up, the car arrives, the doors open, people exit, the mob enters, the doors close (no one else can get on), and you ride. It's the same in the *STELLA* software. Stuff arrives at an oven. If the oven is currently "baking," the stuff waits (in a queue, or a reservoir). When the "baking cycle" is complete, it exits, and the stuff that's waiting, enters (up to the capacity of the oven, or until the "doors open" time expires). That stuff then "bakes" for the length of the oven's "bake time." It's then disgorged. The *Online Help Files* are once again your authoritative source for detail on oven operation. Ovens also are a *Research* version feature.

Verbs

Nouns are wonderful things, but sans verbs...well, you just can't get very far in writing anything meaningful. Verbs represent actions or activities. Unlike nouns, which exist *at a point in time*, verbs exist *over time*. This distinction is the same, for you folks with some familiarity with financial documents, as that between the Balance Sheet and the Income Statement. The Balance Sheet reports on the state of a business *at a point in time*, say, December 31, 2002. The Income Statement reports on what has happened *over a period of time*, say between, January 1, 2002 and December 31, 2002. So, if stocks tell you how things *are* in a system, flows indicate how things *are going*! As flows occur, they *update* the magnitudes of stocks. The only way for the water level in a bathtub to change is for new water to flow in, or for water that's in the tub to flow out. Without flows, conditions within a system would remain *static*. So, it's flows that give us dynamics!

Like stocks, flows can be physical or non-physical in nature. On the physical side we have activities such as: eroding, being born, delivering, dying, producing, in-migrating, discharging, and raining. On the non-physical side we have activities such as: getting angry, building self-confidence, becoming frustrated, praising, cajoling, discussing, arguing, and learning. Notice all the "ing" endings! It is good practice when naming your flows to use the gerund ("ing") form of the verb. Doing so eliminates ambiguity (in particular, confusion with stock concepts) and also better communicates movement.

Consider for example the difference between the words "hiring" and "new hires." In conversation, both are used to refer to the volume of people who have recently joined an organization. But the former is a rate, or "per time," concept, while the latter is an "at a point in time" (i.e., stock) concept. For example, someone might say we have ten "new hires." Those "new hires" could have flowed into the company over, say, a six-month period. In that sense, they constitute an accumulation of people, a stock! But if someone were using the term "hiring," they'd necessarily be talking about an action. We're hiring ten new people between now and the end of the year. So, in naming your flows, in general, try to use "ing" endings wherever possible.

One of the real benefits that come from using the *STELLA* language on an ongoing basis is that the accuracy and clarity of your verbal descriptions will increase. Ambiguities, which can lead to people talking past each other, will diminish. Communication will become much more efficient and effective! And much of this comes from simply being careful about distinguishing between stocks and flows—a distinction that, as we'll soon see, is vital to doing accurate mental (or computer) simulation. Being able to make this distinction is a core *Operational Thinking* skill!

Flows come in fewer flavors than stocks. There are two varieties, and one "wrinkle." Pictures of all three appear in Figure 2-2.

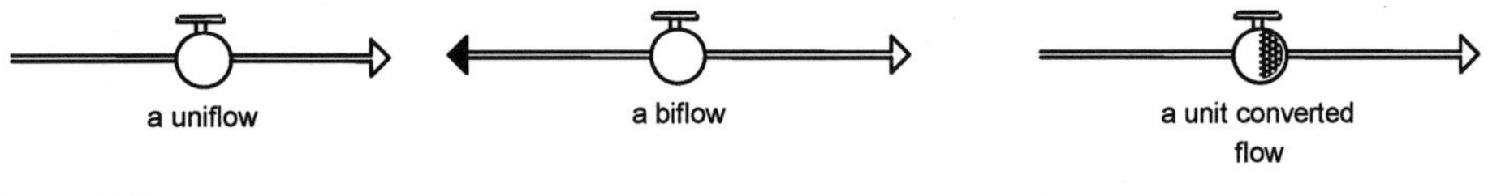

Figure 2-2.
Two Flow Types and One "Wrinkle."

The Uniflow

The standard flow type is called a "uniflow," which is short for "unidirectional." The direction of flow is indicated by the arrowhead. If a uniflow points *into* a stock, it can only *fill* the stock—and vice versa. If a uniflow is an *inflow*, and for whatever reason, its calculated value during a simulation was a *negative* number (indicating that the flow should be *draining* the stock), the calculated value would be over-ridden by a value of *zero*! That is, inflows cannot operate as outflows! Another way to say this is, what you see is what you get! If the diagram shows it as an inflow…that's how it works!.

The Biflow

The other kind of flow is the biflow (for "bi-directional"). It allows flow volume to go in *both* directions, either into or out of a stock. As you'll discover when you learn how to "write sentences," the general rule is that if the processes governing the inflow and outflow to a stock are identical in nature, use a biflow. Otherwise, use a uniflow. A good example of a legitimate biflow is "velocity." If you had a stock called Distance, which represented the total number of kilometers you had traveled away from a starting point in, say, a Northerly direction, the associated *inflow* volume would be northbound velocity. Because you also can turn around (i.e., head Southward), and the process of generating South-bound velocity is identical (except for the direction you are headed in), velocity is correctly depicted as a biflow.

The Unit-converted Flow

We've not yet gotten to the reasons that have motivated the need for including the unit-conversion "wrinkle" in the software. Suffice it to say that, in some instances, it makes sense to convert the units-of-measure of what's flowing, *while it's flowing*! This would enable you to, for example, pull two Hydrogen and one Oxygen atom out of respective stockpiles and make *one*, rather than *three*, water molecules (as illustrated in Figure 2-3).

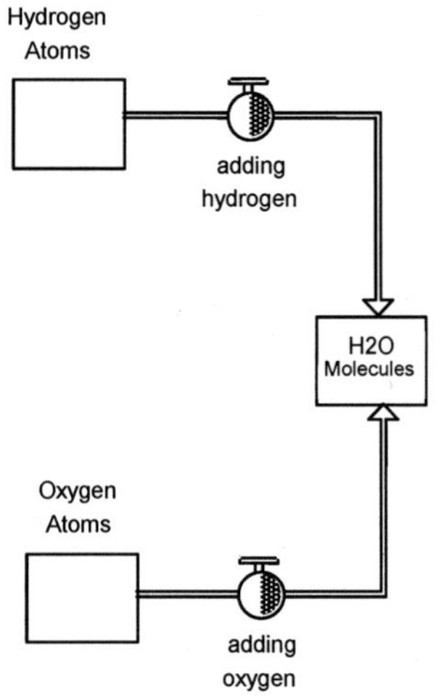

Figure 2-3.
Illustrating Unit-conversion.

In the next chapter, we'll have more to say about the promise and perils of unit conversion. For now, it's sufficient for you to know that it exists as an option in the software.

Distinguishing Between Stocks and Flows

We've made a pretty big deal about the distinction between stocks and flows. But what's all the fuss really about? Why is the distinction so important?

The distinction is so important because stocks and flows constitute the two fundamentally different processes that characterize how reality actually works: *accumulation* and *flow*. If you miss the distinction in constructing your mental models, your mental and/or associated computer simulations are likely to yield erroneous inferences about dynamics! First, I'll illustrate the distinction. Then, I'll illustrate how failing to recognize it can lead to erroneous inferences about dynamics.

Distinguishing Stocks from Flows, a Simple Test

In practice, the best way to distinguish stocks from flows is to perform a simple thought experiment. Imagine that you can instantly "freeze" all activity within a system. This means, in stock and flow terms, that all of the flows instantly become *zero*. Though nothing is now flowing into, or out of, the bathtubs, notice that this does not mean that all the stocks are also instantly zero! Instead, the stocks are frozen at whatever magnitude they were at, at the instant the "freeze" occurred. The magnitudes of stocks *persist*, even if all activity ceases. Let's take a couple of examples to cement the idea.

Suppose you are scolding a child. When you stop, by definition, the scolding activity goes to zero. But the impact of the scolding on the child's level of self-esteem, anger, chagrin, or whatever other non-physical stock the scolding activity may have been feeding or draining, does not go to zero when the scolding stops! The accumulations that have built up, or been depleted, as a result of the scolding activity will be whatever they are when the scolding stops. And, those levels will set in motion a set of coping activities on the part of the child. In effect, the "fun" only begins after the scolding has ceased!

Another example...If we were to cut off manufacturing CFC's (chlorofluorocarbons) tomorrow, the inflow to the stock of CFC's not yet installed in cooling devices would go to zero. But the large stock of CFC's already residing in the earth's lower and upper atmospheres would persist and, while it did, it would continue to deplete our ozone layer for another 90 years or so!

The "So What" Erroneous Inferences

Accumulation and flow are, hence, fundamentally different in nature. And, viva la differènce! It is the existence of stocks that enable flows to vary, sometimes wildly, without causing major disruptions to our lives. Water stockpiles enable communities to withstand droughts. Food reserves guard against a poor growing season. Cash reserves and debt enable keep businesses in business despite negative profits. Inventories allow supply and demand to be out of balance for a while. Without stocks, we'd always be living literally hand-to-mouth. We'd have no buffers or "shock absorbers" to protect against variations in rates of flow—the inevitable "slings and arrows of outrageous fortune."

The fact that accumulation and flow constitute two fundamentally different processes by which reality operates is interesting in an intellectual sense. But "so what?"

The "so what" comes in the form of erroneous inferences that can be drawn from simulating models (be they mental or computer-based) that fail to recognize this important distinction. To illustrate...

Suppose you are a senior Peace Corps official residing at headquarters in Washington DC. You are mulling over two reports from two Country Directors who had been charged with "turning things around" in their respective countries. Country 1 had been experiencing extremely rapid population growth due to astronomically high birth rates. Country 2 had the opposite problem. It had been undergoing a protracted population collapse because of starvation and disease.

The Director from Country 1 reported that recently implemented birth control programs had proven very successful. She provides data that documents a precipitous drop in birth rates that occurred over the last year. The Director from Country 2 reports that recently implemented disease prevention and food relief programs had proven very successful. He provides data that substantiates the precipitous decline in death rates that has occurred over the last year.

Figure 2-4 documents the success of the two Programs, reproducing the plunging birth and death rate data provided by the two, respective Country Directors.

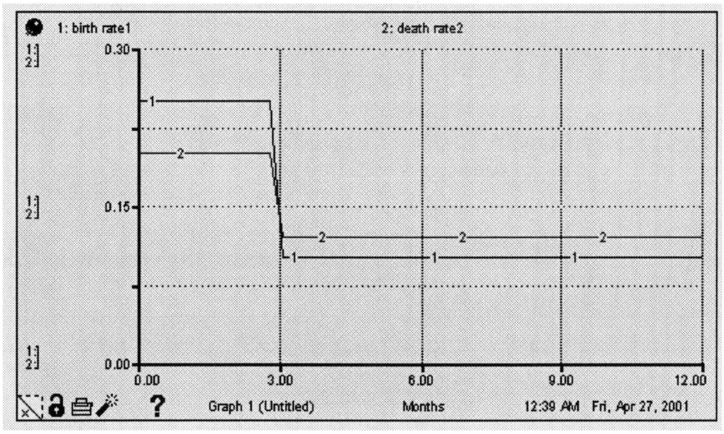

Figure 2-4.
Plunging Birth Rates, Plunging Death Rates.

Are you, as Senior Official overseeing management of efforts in the two countries, happy?

Before you pat yourself, or your Country Directors, too hard on the back, take a look at Figure 2-5, which shows the associated population data from the two countries. As the Figure indicates, Country 1's population growth has been slowed somewhat by the precipitous decline in birth rates—but it is still growing quite rapidly! And the decline of Country 2's population has been slowed somewhat by the precipitous drop in death rate—but it is still declining at a pretty good clip!

What these results illustrate is the difference between causing a *flow* to move in a certain direction, and causing the associated *stock* to move in that *same* direction. Just because you reduce a rate of inflow doesn't mean that the stock fed by that inflow has likewise been reduced. And, just because you have lowered a rate of outflow, doesn't mean that a stock being drained by that outflow also will decline. If you do not recognize the distinction between stocks and flows, you are very likely to miss these kinds of distinctions when attempting to think through dynamics.

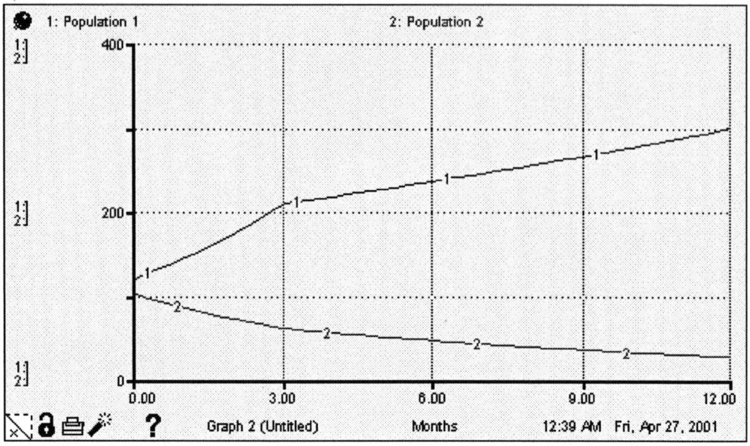

Figure 2-5.
Population Data from Country 1 & 2.

What's Next

You've now "got" nouns and verbs. A good way to practice making the distinction between them is to catch the failure to do so that frequently occurs in newspaper articles, memos, and general discussions. In my experience, for example, numerous arguments have been defused simply by pointing out that one person is focused on the stock, while the other is focused on the flow. As a result, someone will be arguing that conditions are really deplorable, while someone else will be saying we've been making a lot of progress on improving them. And, they'll both be right…but they'll continue to "talk past" each other until the stock/flow distinction is recognized.

As mentioned, being able to distinguish between stocks and flows is a core *Operational Thinking* skill. It's the foundation for all of the other *Operational Thinking* skills—and *Operational Thinking*, in turn, is at the very heart of Systems Thinking. Work on mastering this important distinction! Doing so, will help you to represent elements in your mental models in ways that better reflect how reality actually works! In the next chapter, you'll continue building your *Operational Thinking* skills by learning how to put stocks and flows together to form "sentences."

Chapter 3

Writing Sentences

Operational Thinking

You can't get very far with just nouns and verbs by themselves. You really need to put them together to begin to be able to say something. When you do, you create a "sentence." Sentences, in turn, are the building blocks of paragraphs. Paragraphs are *interesting*! So, learning to write sentences is important.

Defining a Sentence

A "sentence" is a noun with one or more attached verbs. Simple sentences involve only one stock, with associated flow(s). Compound sentences—which later in the Guide we'll refer to as "infrastructures," "spinal cords," and/or "main chains"—involve at least two stocks linked by at least one flow. Figure 3-1 shows a picture of a few simple, and compound, sentences.

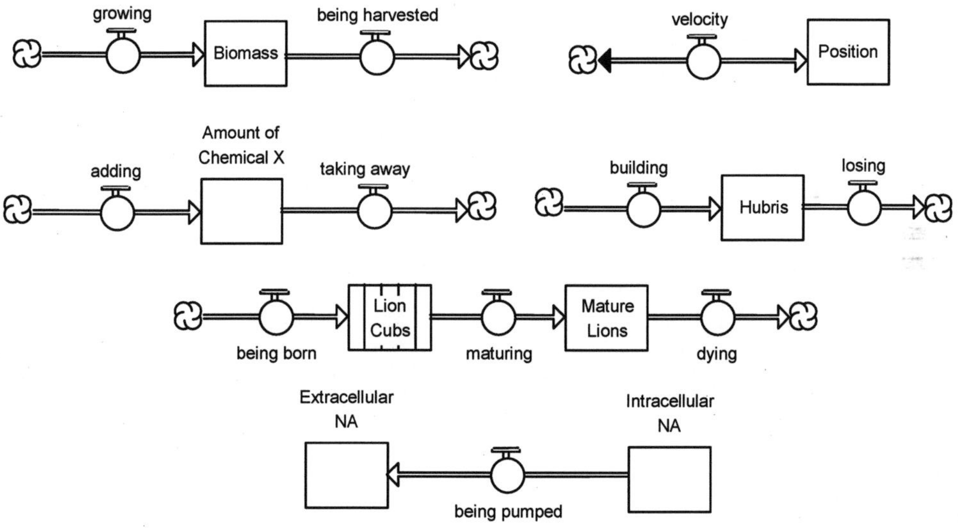

Figure 3-1.
Simple & Compound "Sentences."

Grammar

No, not your father's, or your mother's, mother, but rather the rules for composing sentences. The *STELLA* language like any other language has some of these. Fortunately, it has a very limited number. Two, to be precise. Learn and abide by these rules and you'll be a master *STELLA* sentence-writer!

Rule 1: Respect Unit Consistency

The first rule is to "respect unit consistency." Simply stated, this means that the units-of-measure of the flows attached to a given stock must be the same as the stock's, except for "per time." This rule gets bent a bit when "unit conversion" is invoked—which is why, in Chapter 2, we cautioned against exercising this option too frequently.

In practical terms what this first rule of grammar means is, don't flow toothpaste into a vat of envy. Don't mix apples and oranges. Once you have decided on a unit-of-measure for a stock, always check to make sure that the stuff flowing into and out of it has the *same* units-of-measure (with the addition of "per time").

This first rule seems straightforward enough, but it "goes against" instincts conditioned by Laundry List Thinking (as discussed in Chapter 1). When mental models are constructed using this paradigm, lists of factors "drive" an outcome. There is NO concern whatsoever given to units-of-measure. Arrows, denoting "this impacts/drives/influences that" run from each factor to the outcome variable. Recall the Universal Soil Loss equation and the "Academic Success" examples from Chapter 1.

In Systems Thinking, "impacts/drives/influences" is *not* good enough. Systems Thinkers are striving to capture *causality*—i.e., how things actually work. And, the first step in "representing it like it is," is to recognize the distinction between stocks and flows. The second step is to respect the unit-consistency relationships that exist between the two. Let's take a look at a couple of examples, so you can get a better feeling for the challenges involved. To assert that "*salary levels contribute to employee motivation*" wouldn't be perceived as erroneous. If we were to use the *STELLA* software to "draw a literal picture of the statement," it would look like what you see in Figure 3-2.

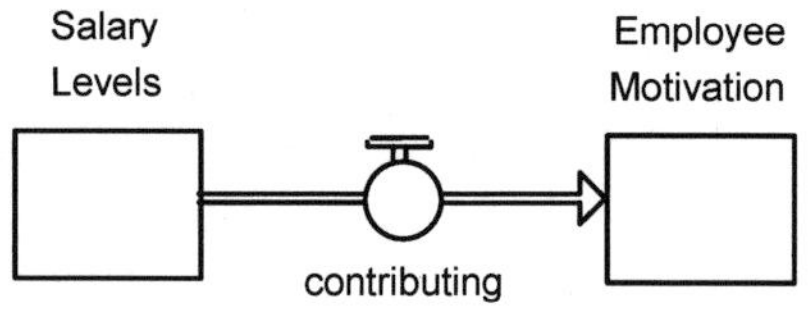

Figure 3-2.
Salary "contributing to" Motivation.

Though plausible-sounding when written in words, if you mentally simulate the structure depicted in Figure 3-2, you would get a crazy result! You'd discover that in order for *Employee Motivation* to *increase*, *Salary Levels* would have to *decrease*! That makes no sense at all! What someone means when they draw a picture like the one shown in the Figure is that *Salary Levels* "contribute to," or are an "input to," *Employee Motivation.* When you make this kind of "loose" statement with the *STELLA* software, you'll get what you deserve—simulation results that are ludicrous! Such results alert you to the fact that your mental model doesn't work the way the real world works. That's important feedback. It help you learn; i.e., to improve your mental models!

It's easy to fix the diagram pictured in Figure 3-2 to make it better describe the real relationship between salary and motivation, but that's not the purpose of this Chapter. Here, what's important is that *STELLA* "sentences" should be accurate descriptions of the relationship between accumulations and the flows that feed and drain them. Flows do not "influence" stocks. Flows do not "have impacts on" stocks; nor are they "inputs to" stocks. Flows *fill* and *drain* stocks; they make the level of stuff in the bathtub go up, and go down. If the bathtub happens to have self-confidence in it, then self-confidence better be what's flowing into, and out of, it. So, for example, it's not "praise" that's flowing in—as one might conclude from the plausible-sounding statement: "praise builds self-confidence." It's okay to say that, just don't draw the *STELLA* picture that way! It would be more accurate to say that praising is one of the "activity bases" for building self-confidence.

Bottom line: when constructing "sentences" using the *STELLA* software, *always* check to ensure unit consistency between a stock and any flows that fill or drain it. If the units do not match (except for the "per time" associated with the flows), you have *not* accurately depicted how that aspect of reality works. And, when you violate the rules of grammar, you will pay the price in terms of unreliability of the associated simulation results!

Rule 2: Respect Conservation Laws

The second rule of grammar actually is pretty closely related to the first. There is a famous "Law" in physics that holds that physical stuff is neither created nor destroyed, but only changes form (i.e., matter and energy are "conserved" quantities). That Law pretty well describes an aspect of how the physical universe works. When you take possession of some chunk of physical stuff, it "came from" somewhere else. It is now absent from that somewhere else. However, because it has changed its location, doesn't mean there is any less of it in existence. When all is said and done, there's still the same *total quantity* of stuff! That's the Law of conservation of matter and energy.

In your *STELLA* models, you will regularly violate this Law. You *must*...otherwise, you'd never be able to set bounds for your model. However, there are legitimate ways to violate it...and illegitimate ways to violate it. Stay "legit" and you'll have no problems with your models. There are two legitimate ways to violate the conservation law...

The first is to make a *conscious decision* to end a particular chain of conserved physical flows. The rationale for doing so is that what you are leaving out of the model is not germane to the issue you are using your model to address. Consider the *STELLA* "simple sentence" depicted in Figure 3-3 as an example...

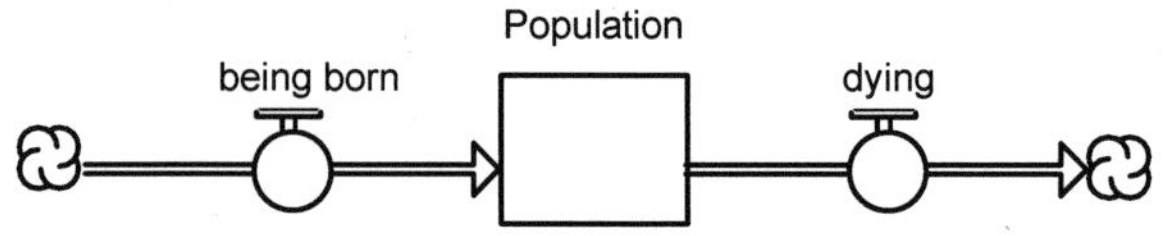

Figure 3-3.
A "Simple Sentence."

The "clouds" at the end of the two flows suggest that people are being born and dying out of, and into, "thin air." Obviously, we are violating the hallowed Law of conservation of matter and energy. But hopefully, we are doing so *consciously*. For example, we *know* people actually come from somewhere else (fertilized eggs). That "somewhere else" is in reality a stock, not a "cloud!" But we are willing to live with the assumption that, *for the purposes the model is to serve*, there are no "important issues" associated with where they come from. For example, we'd be thinking that birth defects are "not an issue" that would be addressed by this model. Any such assumption may be wrong...but at least: (1) the assumption has been made *explicit*, so that others can see/challenge it, and so that you have a constant *visual* reminder that you are making it, and (2) the assumption has been made *consciously*; you're not violating the Law because you're oblivious to it. This is the first legitimate way to violate the conservation Law.

The second legitimate way to violate the Law is whenever you are using a stock to represent a *non-physical quantity*, other than a quantity of time. This is because *non-physical variables do not obey conservation laws!* If you ask the question: Where does knowledge, anger, commitment, or morale, come from? The correct answer is...out of thin air! That's right, no place (and no one) has any *less* knowledge, anger, commitment or morale, because you now have *more*. Everyone can have more of each of these quantities, and no one has to have any less! Non-physical quantities, because they are not subject to physical conservation laws, offer a "free lunch!" And therein lies an important opportunity!

When searching for high-leverage points, a good place to look is in the non-physical domain. That's because, unlike physical stocks, to increase the magnitude of a non-physical stock, you do not have to decrease the magnitude of any other stock. If you re-allocate budget, headcount, or time from one group within an organization to another, one group has less, one will have more. But if you boost the *commitment* of one group within an organization, you do not have to "take" that commitment from any other group! Because non-physical variables do not operate in a zero-sum manner, they are focal points for high-leverage interventions.

As was the case with unit consistency, the notion that non-physical variables do not obey conservation laws seems straightforward. However, many people seem to have trouble with the idea. For example, someone will defend the "sentence" depicted in Figure 3-4 by saying that "customer dissatisfaction leads to employee dissatisfaction." It's hard to argue with the words...but it's easy to argue with the *STELLA* diagram. And it's also easy to resolve any such argument by simulating the model! If you were to simulate this model, you'd discover that when Customer Dissatisfaction goes *down*, Employee Dissatisfaction goes *up*! That is exactly the opposite of what the verbal description implies.

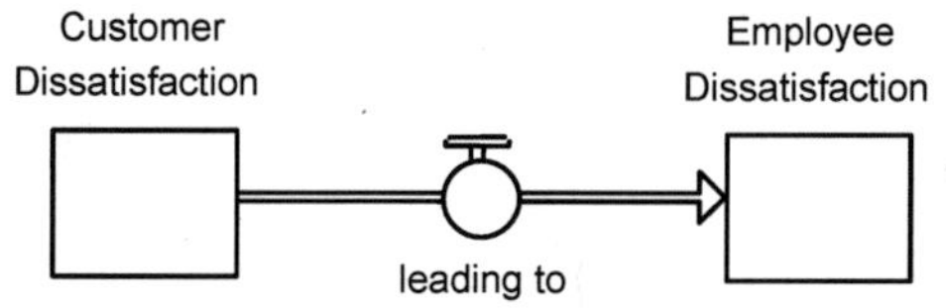

Figure 3-4.
"Customer Dissatisfaction leads to Employee Dissatisfaction."

You can catch the grammatical error by being careful about the units-of-measure, but in this case, it's a little tricky because both stocks are denominated in units of dissatisfaction. So one could argue that there is no unit-consistency problem here. But, there *is* a unit-consistency problem, and the tip-off is the fact that a non-physical quantity (other than time) is being conserved. That's a no-no! And, the simulation confirms it. Customers do not "give" their dissatisfaction to employees. It's not a communicable disease! Through expression of dissatisfaction, customers can stimulate employees to produce feelings of dissatisfaction within themselves. But it is the employees who produce the feelings—customers don't "give them" *their* feelings!

And so, the second rule of sentence construction grammar is: *Do not conserve non-physical quantities* (with the exception of "time"). If you find yourself doing it, check the units-of-measure. You should discover a problem there. If not, run a mental simulation. Ask whether the stock being fed goes up when the stock doing the feeding goes down. If both tests check out, email or call us, we want your example!

What's Next

In this Chapter, you learned how to write a grammatically correct simple and compound sentence. In the next chapter, you'll learn how to link sentences. By the end of that chapter, you'll be well on your way to being able to think *operationally*.

Chapter 4

Linking Sentences

Operational Thinking

On the path to writing short stories, the next important step is to learn how to link sentences together. It turns out there are only two ways to link sentences to each other. Master the distinction between the two, then learn when to use which, and you'll be well on your way to writing rich paragraphs!

Two Ways to Link Sentences

If you enumerate the possible ways to link one sentence to another, you'll discover there are three possibilities…but one of them doesn't work! Figure 4-1 enumerates the possibilities.

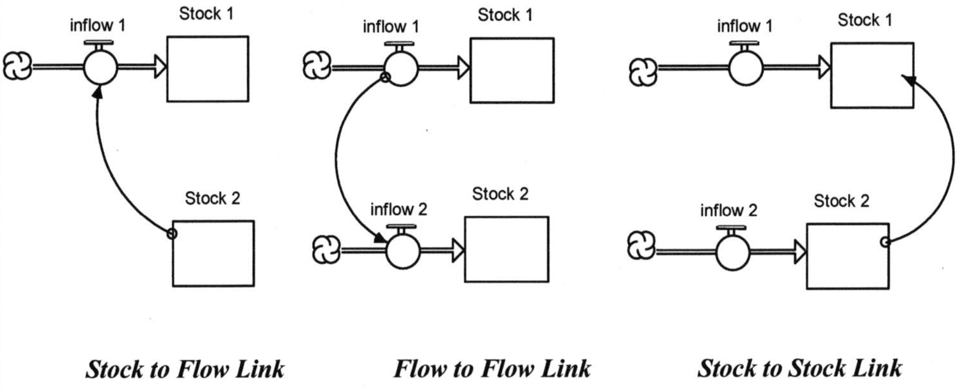

Figure 4-1.
Enumerating Three Possible Ways to Link Sentences.

The first possibility, linking one sentence to another via a connection from a stock in one sentence to a flow in the other, is a possibility. The notion here would be that a "condition" (i.e., the prevailing magnitude of a stock) is generating an inspiration to take some action (i.e., cause the volume of a flow to be greater than zero). A good example would be, say, hunger stimulating you to eat...familiar with that one, are you?

The second possibility shown in Figure 4-1 also is plausible. Here, one action "carries along" another action. A nice example would be your reading of this text and the associated flow of learning that accompanies it...that is happening, right?

The third possibility? Created with smoke and mirrors! The *STELLA* software will not allow such connections! Remember what I said in Chapter 2. The magnitude of stocks can't change by being "influenced by," or "input to." Stock magnitudes change only via filling and draining. Filling and draining are *activities* (i.e., verbs!). And verbs are represented by *flows*, not those skinny little "wires" that you see in Figure 4-1. Only a flow can change a stock. So the only way to link sentences is by linking a stock to a flow, or a flow to a flow. And, as you're about to see, it makes a big difference, dynamically, which one of these two possibilities you choose!

Stock-generated Versus Flow-generated Flows

Stare at Figure 4-2 and decide which of the two representations of the process of "knowledge transfer" that's going on for you with respect to the material you're now reading makes most sense...

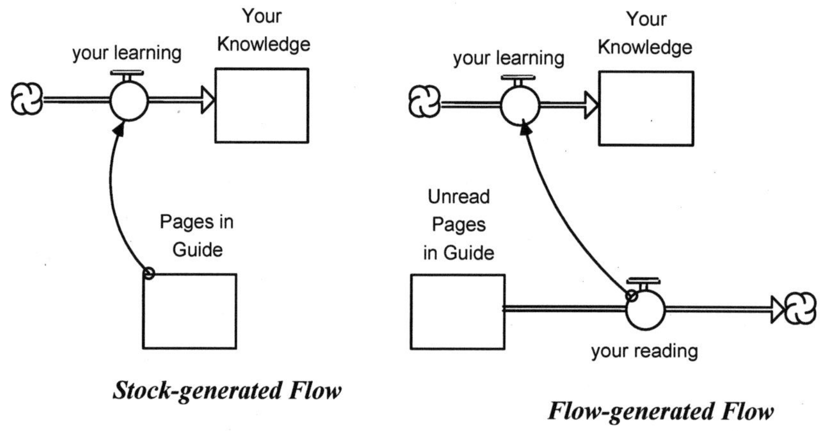

Figure 4-2.
Stock-generated versus Flow-generated Flows.

If you said the representation on the right, you were right! If the representation on the left were correct, all we'd have to do to enable you to learn more would be to add pages to the Guide. You wouldn't have to take any *action* to learn the material contained on those pages, you'd learn simply because the material was there! According to the representation on the right, there is an "activity basis" for your learning. That activity basis is reading. If you stop reading, you stop learning. The latter statement may not be completely accurate because you certainly can learn the material contained in this Guide in ways other than reading it! But, given the simplicity of the representation, it is true that you would stop learning from *that* source (i.e., reading).

Introducing the Connector

In the preceding example, we resolved the issue of which is the better representation by conducting a mental simulation—always a good thing to do, and something the visual nature of the stock/flow language facilitates. However, we also could have simulated the two representations on a computer using the *STELLA* software, and we'd have quickly discovered the problem with the first of the two representations. However you choose to conduct your thought experiments, the first step in linking sentences together is to determine whether it makes most sense to link stock to flow, or flow to flow. And after you've made that determination, you will use "the connector" (the thin wire) to do the linking. Connectors, by virtue of their role as "linkers," become the *conjunctions* in the *STELLA* language.

As you may already have noticed, there are two types of connectors in the *STELLA* language. The one we used in Figure 4-2, the solid wire, is called an "action connector." That's because the wire is transmitting an "action," as opposed to transmitting "information." To make the distinction clear, examine Figure 4-3...

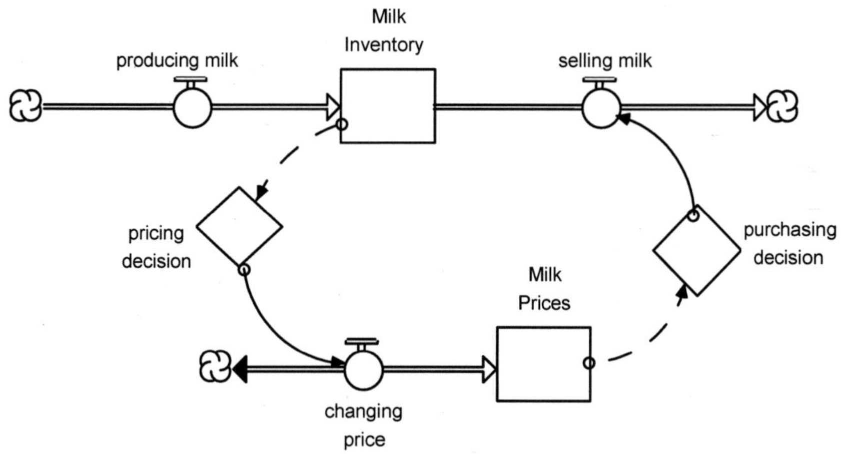

Figure 4-3.
The Supply & Demand for Milk.

In this example, information (represented by the dashed connector) radiates off Milk Inventory levels to serve as one of the inputs to the pricing decision. The decision logic is not visible in the picture because it is embedded within the space-compressed Decision-Process Diamond (DPD) named *pricing decision*. The specifics of that logic are not relevant to the point I'm trying to make here. For information about DPD please refer to the *Help Files* in the software. Out of that decision process comes a pricing decision! It may be to cut price by 10%, or raise it by 20%, or hold it constant. The point is that the information leads to a decision, and the decision, in turn, to an *action*! Thus, the dashed wire begins the process, and the solid wire ends it.

The same is true on the demand side of the ledger. Information about milk prices radiates to consumers. It's part of what influences how much milk they will purchase, and hence how much milk producers will sell. Consumers make their purchasing decisions and then take *action*—i.e., they purchase a certain quantity of milk that day/week/month.

Information connectors thus provide the information that's used to arrive at a decision. Action wires, in effect, convert the resulting decision into an action that ultimately manifests as a change in the volume of flow. The distinct difference in purpose between the two types of connector explains why only information connectors can "stick into" DPD's. However, both types of wire can "come out" of a DPD because in addition to the action that will be taken as a result of the decision, information about the decision, or the inputs to that decision, also can be transmitted.

Information and Action connectors are similar in that neither can be used to represent a conserved-flow linkage. That is to say, no "stuff" flows through either type of wire! When information is "radiated," there isn't any less of it left to radiate! Thus, for example, when you step on the bathroom scale, and information about your body weight radiates off the dial, no actual pounds are being lost through that radiation. It's not body weight that's radiating, it's *information about* body weight that's radiating!

And so, flows transport; wires transmit. Connectors serve as "inputs" and "outputs," not "inflows" and "outflows." Being able to grasp these distinctions is another of the sub-skills that comprise *Operational Thinking*.

The astute observer would have noticed a little problem way back in Figure 4-2. If you'd like, mosey on back there and see if something about the representations in that Figure bother you. I'll wait...

Were you bothered by the fact that all of that unit-consistency brouhaha that I threw at you back in Chapter 3 seemed to fly out the window? Well, you should have been! Let's focus on the second representation in Figure 4-2. We said this was the more accurate of the two depictions. What are the units-of-measure of the *your reading* flow? If you're having trouble with the question, remember that the units of a flow must be the same units as the stock to which it's attached, except for "per time." The stock is denominated in "pages." Therefore, *your reading* must be dimensioned as "pages per time."

There's an action wire that runs from the *your reading* flow to the *your learning* flow. What are the units of the latter flow? Again, you may wish to begin with the stock to which the *your learning* flow is attached and work backward. Hopefully, you concluded that the units of the flow must be "knowledge per time" (or "understanding per time," or some such). But how can that be, if the wire coming *into* the flow from *your reading* does not have those units-of-measure? The answer is: *it can't*! We need another concept here, folks. And it's not just so we can make the units work out right. It's so we can make the representation *operational*: i.e more accurately reflect the way reality works! Insisting on unit-consistency is not just an anal-compulsivity to which Systems Thinkers have gotten addicted. It's a way to ensure that your representations better reflect how things really work!

In this case, let's discover the missing concept by thinking about the process—rather than backing into it by figuring out what the "units" need to be in order to cause the *your learning* flow to have the correct units-of-measure.

Here's a thought experiment: If a seven year-old child were to read these pages that you have been reading, would they have learned as much as you have learned? Unlikely. Why? Because, presumably, your cumulative learning experiences have made you both a better reader than a seven year-old, and you also have amassed more content and understanding (i.e., you have more "hooks") with which to pluck understanding from the words and pictures on the pages you're turning. In addition, you are likely to be more motivated to learn this material than the average seven year-old. All of these factors will conspire to cause you to learn more "per page turned" than a seven year-old. Operationally speaking, your "learning productivity" (units-of-measure: "learning per page") is higher! If we add "learning productivity" to the picture, we end up with Figure 4-4.

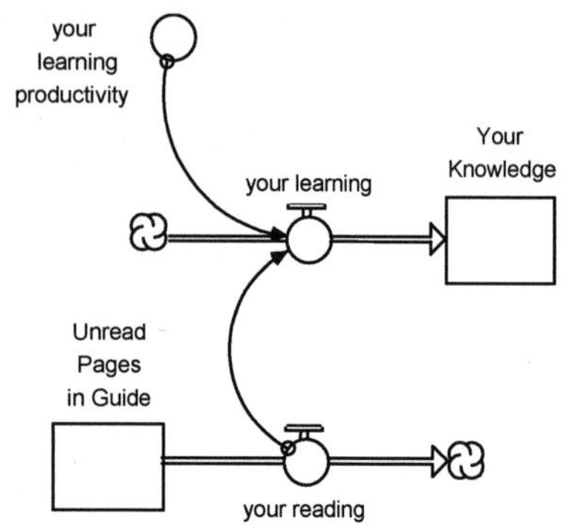

Figure 4-4.
Correcting Figure 4-2.

Introducing the Converter

The *STELLA* language element that we used to represent *your learning productivity*, and that often is used to represent "productivity" in one of its many incarnations, is called a *converter*. In this context, the converter is playing the role of an "adverb." It is modifying the verb *your learning*. It tells how much learning occurs for a given unit of the "driving activity" (in this case, *your reading*). From a unit-consistency standpoint, it "converts" the units brought into the learning flow from the reading flow (i.e., pages/time) into the proper units of learning (knowledge/time). If you would like to scrutinize the algebra, it would look like this:

your learning	=	**your reading**	x	**your learning productivity**
(knowledge/time)		***(pages/time)***		***(knowledge/page)***

Note that (pages/time) times (knowledge/page) is equal to (knowledge/time), so that the units-of-measure on the left-hand side of the equation balance with those on the right-hand side of the equation—this makes life good, physicists smile, and algebra teachers jump for joy. It also yields representations that more accurately mirror how reality works, so that when you simulate those representations for purposes of drawing conclusions, you have a greater chance of being able to rely on those conclusions. *Operational Thinking* rules!

And so, converters often play the role of "adverbs," modifying flows. In this role, they tell how much of a contribution to an activity is being made per unit of the "driver" of that activity—be that "driver" a flow (as in the example we just examined), or a stock. Let's look at two more examples, just to cement the concept. Examine Figure 4-5…

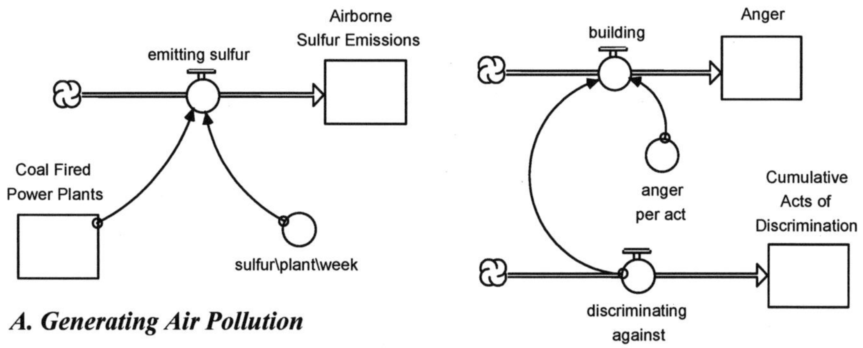

Figure 4-5.
Stock & Flow-generated Flows.

The first representation in Figure 4-5 shows the generation of sulfur emissions from coal-fired power plants. The *emitting sulfur* flow is being represented as a *stock-generated* flow. In the second representation, the *building* of anger is a *flow-generated* flow. In both examples, converters are used as "productivity" terms.

The converter, *sulfur\plant\week,* is being used to convert the number of *Coal Fired Power Plants* that are operating at any point in time into a flow of *sulfur emissions*. It has the units: sulfur per plant per week.

In the *Anger* example, a flow of being discriminated against is driving the buildup of anger. The productivity term, *anger per act*, indicates how much anger each act of discrimination generates—which is to say, how "productive" each act of discrimination is in causing anger to build up. Its units of measure are: anger per act of discrimination.

The preceding examples should help to drive home the concept of converters as "adverbs," or more operationally couched, as "productivity terms." We're on solid ground here, both grammatically and conceptually, in terms of describing how many processes actually work. In fact, the two flow formulations illustrated in Figure 4-5 reoccur so frequently in *STELLA* models that we've given them generic names. The *stock-generated* formulation is called an "External Resource Template," and the *flow-generated* formulation is called a "Co-flow Template." You should study these two formulations. You'll find them to be extremely useful in constructing models using the *STELLA* software. In Chapter 5, we'll introduce three more such *generic flow templates* for a total of five. I make use of one of these five templates to specify 90% of the flows in the models I construct. Being able to creatively adapt and employ these templates is the hallmark of someone who has mastered *Operational Thinking*.

The flow templates are things of beauty. But now, we're going to balance all this pulchritude with a little "ugliness"...

Converters as Pandora's Box

It turns out that those nice, innocuous-looking little circles we call converters can function as more than just adverbs. They can operate as adjectives, dangling participles,...and just about any other dern thing! As they say in New Hampshire..."Ahyup, there's flies in that there ointment." Converters become a catchall for: doing algebraic operations (like summing or division), representing exogenous inputs, and serving as substitutes for either stocks or flows that you are choosing (for reasons of simplification) to not represent as such. I'll briefly illustrate a few of these practical, yet not so beautiful, uses of converters here. You then can probably discover even more uses by perusing the various models that come with the *STELLA* software.

Several non-adverbial uses of converters are illustrated in Figure 4-6.

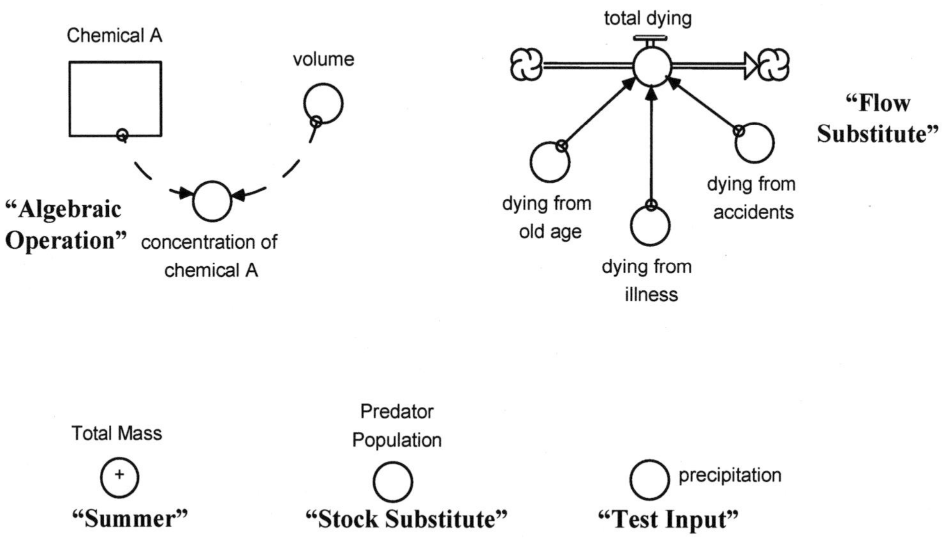

Figure 4-6.
Illustrating Some Non-adverbial Uses of Converters.

Algebraic Operations

The use of a converter for performing an algebraic operation is the first of the uses illustrated in Figure 4-6. In the example, the concentration of a chemical is being calculated—a simple division of the quantity of the chemical by the volume within which that quantity is contained. Calculating density would be another good example of this use of a converter. Often you will wish to "bury" such calculations inside a Decision-Process Diamond (DPD). This "gets the algebra" off center stage—which should be reserved for stock/flow plumbing and feedback loops (discussed in the next chapter).

Summer Converter

Summer is something we dearly wished we had more of in New Hampshire. So, we added it to the software. And, in addition to serving as a constant reminder of what we don't have, it also is useful for "adding up" quantities without having to "run all the wires" into a poor hapless converter. Summer converters, a choice within the converter dialog box, allow you to add up any quantities you like just by clicking on them in the Allowable list. You'll find it to be a useful bit of functionality on a number of occasions.

A Stock Substitute

Predator Population, the variable chosen to illustrate "stock substitute," is in concept a stock. However, if you are not "interested in" the inflow to the stock (i.e., *being born*), or the outflow (i.e., *dying*), you may want to simplify things by just representing the stock as a converter. As we'll see in a later Chapter, converters *can* change over time. They are not always just constants! So, using a converter to substitute for something that's really a stock, doesn't mean you lose the ability for that variable to change with time. It just means that you will consider changes as relationships not included within the model boundary. More on all of this when we get to "feedback loops" in the next chapter. For now, suffice it to say that there are instances where simplification dictates that you represent something that is, in concept, a stock, with a converter.

Flow Substitute

The *total dying* flow in Figure 4-6 illustrates the use of converters to substitute for what are, in concept, flows. No problem. Rather than having the three types of dying represented as flows, you can represent them as converters and then, as in the illustration, sum them up into a *single* flow. There is one issue you should be aware of when you do this. When numerical values are reported in Tables within the *STELLA* software, variables represented as converters are calculated *before* variables represented as flows. So, when using a converter to represent a flow, in order to cause the numerical values as reported in Tables to appear at the same time as flows do, it is necessary to click the C→F button (in the Table dialog box) after entering the converter into the Selected list. You can read more about this in the *Help Files*.

Exogenous, or Test, Input

As you've seen if you've been through the software tutorials, or just played around with the software, in each converter's dialog box, there is a scrollable list of "Builtin" functions. These "functions" enable you to create various kinds of "patterns" (like ramps, steps, randomness, sinusoids, etc.) that are useful for "testing" your model and, in so doing, serving as "exogenous inputs." We'll have more to say about these variables when we discuss model testing in Chapter 9.

What's Next

Over the last couple of chapters, you have become intimately acquainted with the essence of what constitutes *Operational*

Thinking—a major Systems Thinking skill. A second major skill is called *Closed-loop Thinking*. Put these two "biggees" together, and you can apply for a Systems Thinker's union card. You'll also be able to write good paragraphs—the building blocks of short stories.

Chapter 5

Constructing Simple Paragraphs

Closed-loop Thinking

A "paragraph," in Systems Thinking parlance, is a "feedback loop"—a closed-loop of causality. Previous chapters have alluded to the fact that "paragraphs" are "interesting." Why are they interesting, you may wonder? They're interesting because, like those little wind-up toys…you prime them, and they then take off on their own! Feedback loops *self-generate* behavior. If you bump one…get out of the way!

There are two types of feedback loops: *counteracting* and *reinforcing* (sometimes referred to, by technical people, as *negative* and *positive*, respectively). We'll begin with a formal definition of feedback loops, and then discuss the counteracting and reinforcing variety. In this Chapter, we'll deal only with *simple* feedback loops—where "simple" has a technical definition, and is not simply a measure of associated complexity. In Chapter 6, we'll treat non-simple paragraphs.

Definition of a Feedback Loop

A *feedback loop* exists whenever a "noun" (stock) is linked to a "verb" (flow) in the *same* sentence. The link may be direct, or part of a chain of links passing through other "sentences" first. An example of a direct, and an extended-link, feedback loop appear in Figure 5-1.

In the "direct-link" example (A), a *Bacterial Population* reproduces itself. In the "extended-link" illustration (B), oppression breeds resistance, which fuels more oppression—a nasty spiral that we see operating in prisons and under dictatorial government regimes. In both examples, the noun connects to its "sentence-mate" verb. In the first, the connection is *direct*. In the second, the connection passes through another sentence first. Whenever a noun links back to its sentence-mate verb, a "feedback loop" exists. Feedback loops are extremely important to the functioning of *all* natural, physical, and social systems. Without feedback loops, there would be no life of any kind! These loops are thus pretty fundamental critters. It's worth your while to master the *Closed-loop Thinking* skills needed to understand how they work!

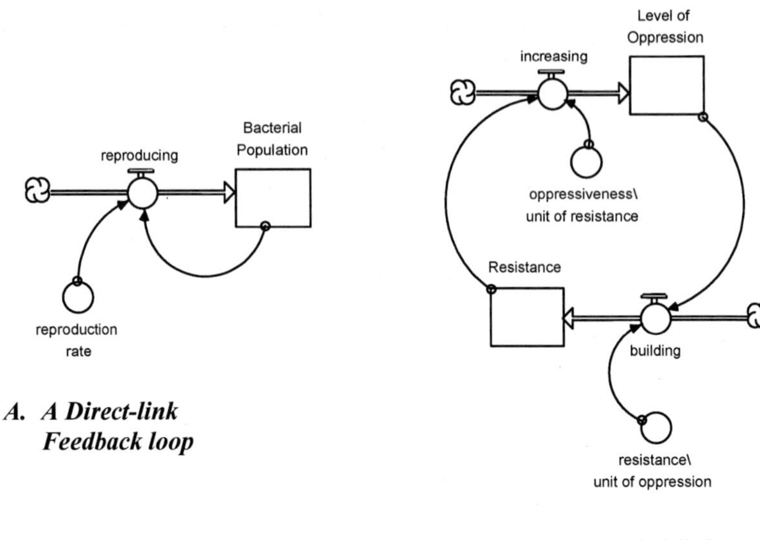

Figure 5-1.
Direct and Extended-link Feedback Loops.

"Simple" Feedback Loops

In building understanding, it usually makes sense to start simple. That's certainly the case with feedback loops, where things can get pretty wild, pretty fast. It's important to have a solid grounding in the basics of the structure and behavior of feedback loops before launching off into building models. For this reason, I have defined what I call a "simple" feedback loop, distinguishing it from a "complex" loop.

A "simple" feedback loop is one that satisfies two conditions. It is composed of a *direct* link (i.e., the stock links directly to its associated inflow or outflow). And, its parameters (i.e., "productivity terms") are *constant*. The example shown in Figure 5-1A is a "simple" feedback loop, while that in 5-1B is not.

Simple Counteracting Feedback Loops

Counteracting feedback loops are so-named because they *counteract* change. Try to push in one direction on something that's being controlled by a counteracting feedback loop, and you'll experience resistance or "push back" in the *other* direction.

Counteracting feedback loops are everywhere! Each cell in your body uses them to maintain the delicate chemical and electrical balances you need to remain alive. Countries use them to maintain trade and arms balances. And every life form in between uses them to maintain order, to keep things in proper proportion. Counteracting loops act to maintain stability. Without some stability, neither life itself, nor growth, is possible! Examples of counteracting loops in action are everywhere...

Implementing change within an organization usually stimulates counter-pressures to slow or undo it. Raising your body temperature by exercising, triggers sweating—a process that works to cool you back down. Falling prices motivate consumers to shop, which depletes inventories and drives prices back up. Committing a faux pas that damages an important relationship stimulates actions to repair that relationship. A buildup of moisture in a cloud inspires precipitation that drains the moisture.

Figure 5-2 depicts the two incarnations of a *simple* counteracting feedback loop. Like the "external resource" and "co-flowing" templates introduced in Chapter 4, the two flow processes shown in Figure 5-2 commonly recur. So, like their predecessors, we have given these templates names. We call them the "draining" and "stock-adjustment" templates, respectively. An Appendix at the end of this Chapter summarizes the five generic flow templates that we have identified in our work (you'll be introduced to the fifth later in this Chapter). As stated in Chapter 4, we use one of these five templates to specify 90% of the flows in the models we construct. If your intention is to become proficient in applying Systems Thinking, time spent mastering the structure of these templates, and when it's appropriate to use each, is time *extremely* well spent!

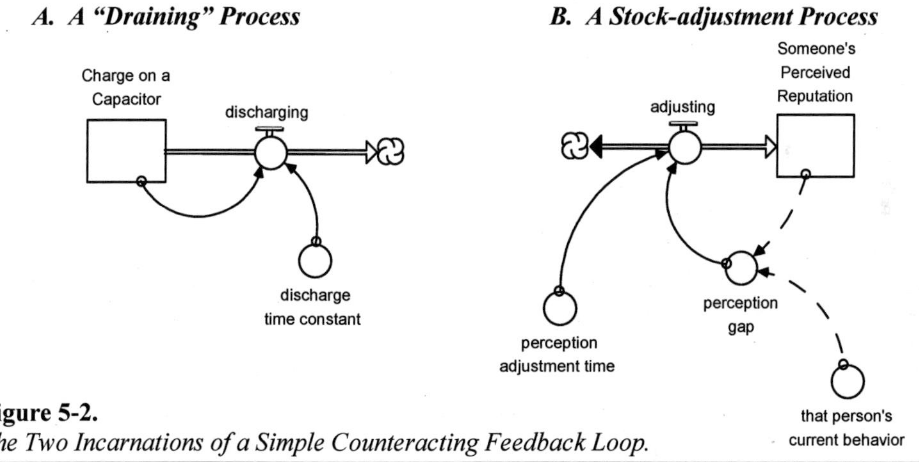

Figure 5-2.
The Two Incarnations of a Simple Counteracting Feedback Loop.

The Draining Template

The "draining" template is used primarily to capture *passive decay processes*. In the example shown in Figure 5-2, the charge on a capacitor is "decaying." Pull the plug on your laptop's power supply. Notice that the little green "on" light doesn't go off instantly, but kind of fades off. That's because it takes some time for the charge that is being stored in the capacitors to decay. Other common examples of draining processes include: the decay of any kind of awareness, memory loss; fading perceptions; water running down a drain, after you open the drain; and, heat energy dissipating out of a steaming hot cup of coffee left sitting on a counter.

Draining processes are so named because of how they behave when "unopposed"—i.e., when they have no inflow to the associated stock to offset them. Under these circumstances, draining processes *drain* stocks! The pattern the magnitude of the associated stock traces, when the "draining fraction" or "draining time constant" is constant, looks like what you see in Figure 5-3A. Initially, the charge on the capacitor is held constant because the discharging rate is being zeroed out. Then, the "zeroing out" process is neutralized, and the capacitor is allowed to freely discharge.

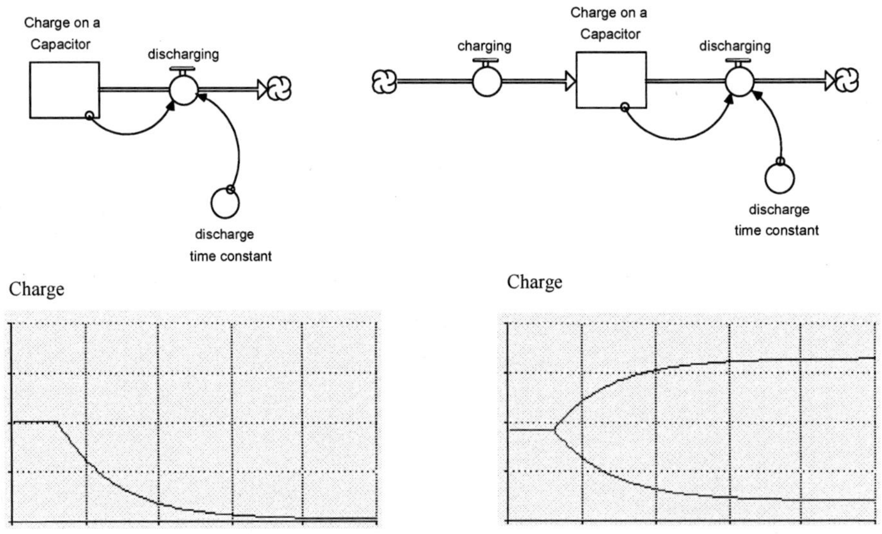

A. Draining Process, "Unopposed"

B. Draining Process, "Opposed"

Figure 5-3.
The Behavioral Repertoire of the Draining Process with a Constant Draining Fraction.

The resulting pattern of decay is known, mathematically, as a "negative exponential." But because we believe in remaining "positive," we'll just call it "exponential decay." This pattern also is colloquially referred to as "half the distance to the wall." To see why, imagine that the stock involved is *Distance from the Wall* (measured in meters). The draining flow might be called *stepping*, and is measured in units of *meters/second.* The draining parameter might represent the fraction *per time* of the stock "drained." Let's assume that fraction to be 0.5 (or 50% of the magnitude of the stock per second; assume you take 1 step per second to keep things simple). Let's say you are initially 3 meters from the wall. In your first step, you'd drain 1.5 meters from the stock—you've gone ½ the distance to the wall. You are now standing 1.5 meters from the wall. On your next step, you drain ½ of what is left in the stock, or 0.75 meters. And, so on. Each step you take will eliminate ½ of the remaining distance to the wall. Hence, the name.

The astute observer will quickly determine that if someone were actually executing the "half the distance to the wall" experiment, they would *never* quite reach the wall. This is true, but it turns out that in "three times the 'time constant'" (the "time constant" being defined as the reciprocal of the "draining fraction"), the magnitude of the stock will be "close enough" to be considered "there" (about 95% of the initial magnitude will have been drained).

Figure 5-3B shows how a draining process behaves when "opposed" by an inflow. The two lines on the graph reflect the two simulations that were conducted. For a brief period, at the outset of both simulations, the inflow to, and outflow from, the stock are constant and equal. The magnitude of the stock is therefore unchanging. That's why the two lines initially are flat and equal (and that's also why you see only *one* line initially). In the first of the two simulations, the inflow steps up to a *higher* constant volume. In the second, the inflow steps down (by the same amount) to a *lower* constant volume. As you can see, the pattern of charge on the capacitor traces symmetrically opposite curves.

In the step-down case, the draining process manifests in the "classic," exponential decay pattern—except that rather than the stock draining all the way down to zero (as it did when the draining outflow operated unopposed), it decays toward a *non-zero* level. What that level will be can easily be calculated. The stock will stop falling when the outflow volume has decreased to the point where it is once again equal to the stepped-down inflow volume. At that point, inflow and outflow are equal. Hence, the magnitude of the stock will remain constant. The outflow volume is calculated by multiplying the current magnitude of the stock by the draining fraction. When this magnitude has declined to the point where the multiplication produces a value equal to the inflow volume, the decline will cease.

In the step-up case, the draining process doesn't manifest its presence in the classic, exponential decay form. In fact, there's no "decay" at all! But there is a mirror image, exponential process at work. This upward "half the distance to the wall" pattern is known as "asymptotic growth." What's going on in this case is that because the inflow has been stepped-up (above the initially constant outflow), the magnitude of the stock begins to grow. As it does, the outflow volume (which, again, is calculated by multiplying the magnitude of the stock by the draining fraction) begins to increase. As the outflow volume swells, the magnitude of the stock is continuing to grow—but ever more slowly. When the outflow volume increases to the point where it equals the stepped-up inflow volume, the magnitude of the stock will cease increasing, and the system will once again be back in steady-state.

So, as you've seen, what is called a "draining process," doesn't always manifest that way. It's more accurate to describe it as a member of the "half the distance to the wall" processes—named for the characteristic pattern of behavior exhibited by *counteracting feedback loops* operating in isolation. As you are about to see, the draining process is just a special case of the other member of the set of such processes—the *stock-adjustment process*.

The Stock-adjustment Template

Figure 5-4 portrays the Stock-adjustment template and its associated characteristic behavior patterns.

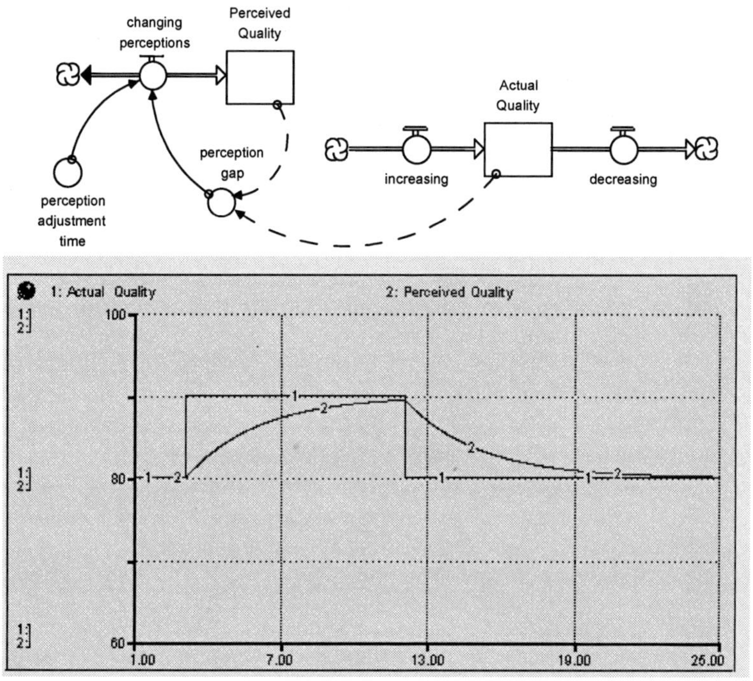

Figure 5-4.
Stock-adjustment Template and its Associated Characteristic Behavior.

If you take a quick peek back at Figure 5-3B, you will see the similarity in the patterns of behavior generated by an "opposed" draining process and a stock-adjustment process. Both generate "half the distance to the wall" patterns. As stated previously, the draining process is simply a "special case" of the stock-adjustment process. Specifically, it's a stock-adjustment process in which the "goal" toward which the stock is adjusting (in the example, *Actual Quality*) is never larger in magnitude than the stock—which is to say, a stock-adjustment process with the flow only flowing *out* of the stock.

Counteracting Loops: In Summary

The bottom line on *simple*, counteracting feedback loops is that they exhibit "half the distance to the wall" behavior patterns. They either decay exponentially toward some goal (or target magnitude), or they increase asymptotically toward a goal. The story of counteracting feedback loops becomes a lot more interesting when we extend the links to form loops involving more than one "sentence," and also when we allow the associated parameters (draining fractions and perception adjustment times) to vary. But these "more interesting paragraphs" are for Chapter 6. Let's now look at the simple *reinforcing* feedback loop.

Reinforcing Feedback Loops

Reinforcing feedback loops are so-named because they *reinforce* change. Push on something that's being controlled by a reinforcing feedback loop, and you'll start an avalanche!

Reinforcing loops are less prevalent, in both natural and human-populated systems, than their counteracting brethren. And that's fortunate. When you mess with a reinforcing loop, you've got a tiger by the tail! Tigers are powerful. Harness the power, and you have a wonderful engine for growth, change, or evolution. Lose control of the power, and you have a powerful engine of destruction! Here are a few examples of reinforcing loops in action…

The surge in popularity following the introduction of a new, "hot" website, music CD, or movie. The meteoric run-up, and subsequent free-fall, in stock prices during the dot.com boom/bust. The rapid proliferation of cells in a cancerous tumor. The spread of an infectious disease or a new fad through a population. Road rage. The "recruiting" of resistors and zealots, against and for, an organizational change initiative. The skyrocketing of free agent salaries in major league sports. The mushrooming of population in US sunbelt cities. All of these examples illustrate reinforcing feedback loops at work. Reinforcing loops "feed upon themselves." They are "compounding in nature. There is nothing inherently "bad" about such processes. But, and this is an important "but," *no* such process can continue forever!

Anything that feeds upon itself must ultimately reach a limit. Either the limit is "self-imposed," or it will be "externally-imposed." In the self-imposed case, actors within the system decide that "enough is enough," and take some action to defuse the compounding engine. So, for example, when a price war erupts between competitors, or an arms race between countries, someone usually pops up amidst the furor and says, "Okay, time out! Things are getting out of hand." Cooler heads prevail, and the compounding process is defused.

In cases where no self-imposed limitations emerge, the environment within which the growth is occurring eventually will "speak." First it will be in a whisper. Then a normal voice. Ultimately its shriek will grow increasingly shrill until eardrums burst and growth *must* stop. Cancer cells, for example, generally don't hear very well. They ignore normal growth covenants imposed by a tissue context. They continue to pile up until available nutrient flows are exhausted and metabolic waste removal capacities are exceeded. The result is that cells, usually at the core of the tumor, begin dying of malnutrition and metabolic waste poisoning. These deaths then offset (and sometimes more than offset) the "births" that are occurring in the outer layers of the tumor, and hence growth comes to a halt.

Self-imposed, or environment-imposed...either way, ultimately enough is enough! Exponential growth *must* cease at some point.

Compounding Template

Figure 5-5 illustrates the *simple* reinforcing feedback loop structure and its associated characteristic behavior pattern. It's called a "compounding" process, and constitutes the fifth and final *generic flow template*. The associated, highly recognizable, pattern of behavior is called "exponential growth." Here, rather than traveling "half the distance to the wall," the wall itself is being pushed away (at an ever-increasing rate). The process is analogous to trying to catch your shadow. The faster you run after it, the faster it recedes from your grasp.

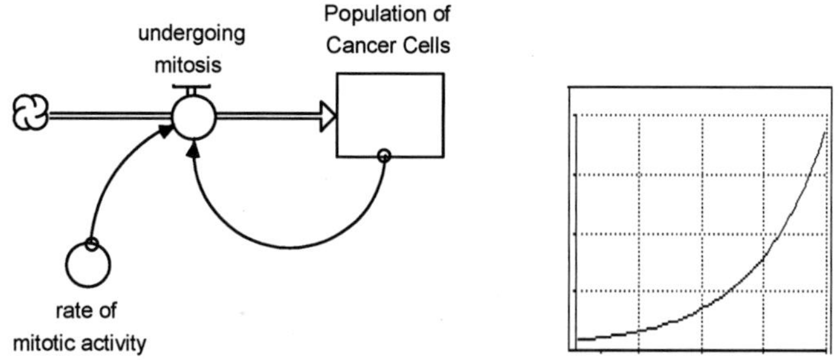

Figure 5-5.
The Structure & Behavior of a Simple Reinforcing Feedback Loop.

The explanation for the pattern of behavior is easy to understand. Stuff flows into a stock, be it money, enthusiasm, or cancerous cells. Once in the stock, it causes even more of like-itself-stuff to flow in. In the case of money, what you have in your savings account is the basis for generating an inflow of interest payments. The volume of interest you earn is proportional to the amount of money that's currently in your account. The growth constant in this case is called the "interest rate." Similarly, for enthusiasm or cancer cells, you get some, and they then bring in more of same. Enthusiasm is "infectious," and cancer cells divide to produce new cancer cells. Either way, the stock "feeds upon itself!" The "feeding upon" process produces a pattern of growth in which each increment of inflow is a constant percentage of the preceding magnitude of the associated stock. As a result, as the stock's magnitude grows, so too does the associated inflow volume—and, by a proportional amount. Hence, the curve of the stock's magnitude (and the flow's volume, as well), takes off, hooking upward as it goes. There's a little story that nicely illustrates the nature of an exponential growth process.

A farmer had a pond stocked with catfish. One fine spring morning, he noticed that a lily pad had appeared on the pond. The next day, he noted that a second pad had come into being. On day 3, there were four pads. After 30 days, lily pads covered one-half the pond. The farmer was concerned about allowing the population of Lily plants to grow too much larger, for fear it would endanger the catfish population. He wondered how much longer he ought to wait before taking some action to stem the growth of the lily pad colony. Can you estimate how many more days it would be prudent for the farmer to wait?

The answer is that the farmer probably already has waited too long. With the lily population doubling *every day*, it will take only one more day for the pond to be *completely covered* with lily pads! That's the nature of exponential growth…it sneaks up on you. And, before you know it, you're toast! Again, fortunately, not many reinforcing loops exist independently of counteracting feedback loops that keep them in check. But because some reinforcing loops are very strong, they can spiral out of control before *consciously-chosen* counteracting loops have an opportunity to kick in. When this happens, it is almost never "pretty." "Environment-imposed" counteracting loops are usually quite unforgiving!

Combining Loops

As already noted, things get considerably more intriguing when we move beyond "simple" loops. But just to complete the "simple" story, and so it's easier to see what "complex" brings us, Figure 5-6 combines a simple counteracting and simple reinforcing loop. The Figure also shows the behavioral repertoire associated with the combination.

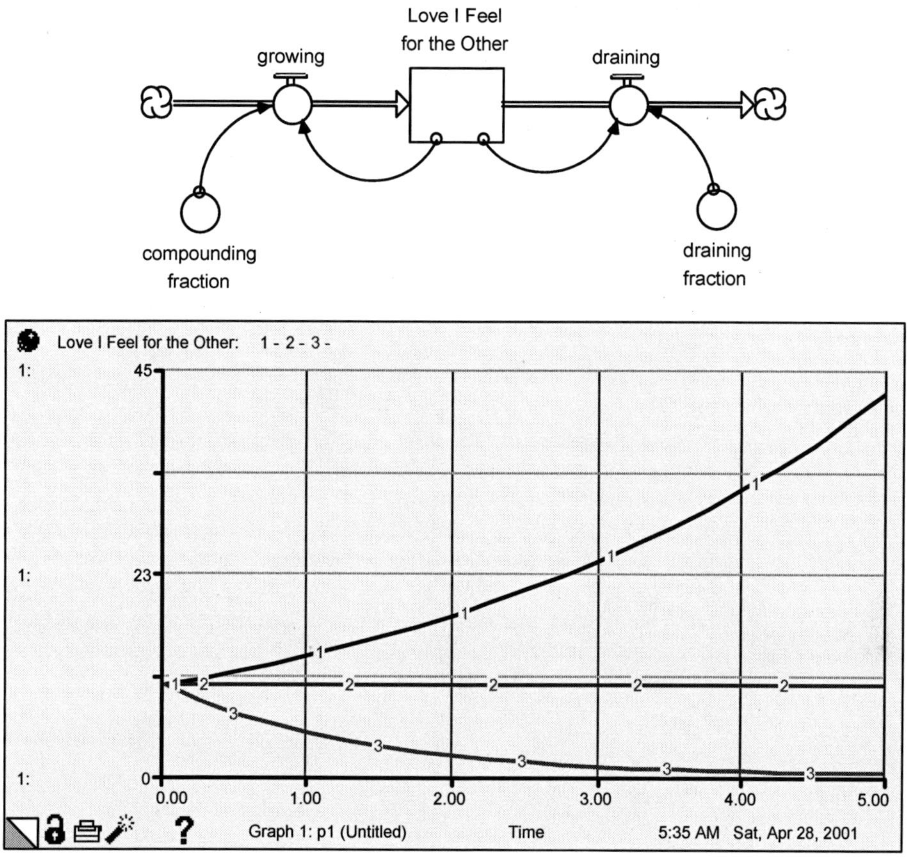

Figure 5-6.
Combining a Simple Reinforcing & Counteracting Loop.

As you can see, when you allow a simple counteracting and reinforcing loop to interact, three patterns of dynamic behavior can be produced—depending on the values of the two parameters (generically named, the "draining" and "compounding" fractions). When the two parameters are equal, the magnitude of the stock remains unchanged; i.e., *neither* loop dominates because they are exactly equal in strength, (line 2 on the graph). When the "compounding fraction" exceeds the "draining fraction," the magnitude of the stock exhibits exponential growth, (line 1 on the graph). This means the reinforcing loop is dominant, and because the parameters are constant, will remain so *forever*. If the "draining fraction" is larger than the "compounding fraction," the counteracting loop dominates and the stock will decay exponentially, forever (line 3 on the graph).

That's it. Not a very elaborate, or very interesting, repertoire of dynamic behavior patterns, is it? Once the two parameters are assigned values, one of three possible patterns of dynamic behavior will emerge and then persist. Chapter 6 examines the consequences of allowing the parameter values associated with feedback loops to change dynamically. What we'll discover is that such changes can cause feedback loop dominance to *shift* over time. For example, a reinforcing loop might dominate in the early going, but the strength of an associated counteracting loop could be building all the while. At some point, this will allow the counteracting loop to overpower the reinforcing spiral (cooling it off!). Such shifts in feedback loop dominance are what create the "non-linear behavior" discussed in Chapter 1, and why *Non-linear Thinking* is such an important Systems Thinking skill to master.

What's Next

OK, you've come a long way, and you have only one more chapter to process in order to complete the "building blocks of short stories" progression that began in Chapter 2. You've been exposed to *Operational Thinking* and also much of what *Closed-loop Thinking* is about. In Chapter 6, you'll finish off *Closed-loop Thinking* and also learn something about *Non-linear Thinking.* Specifically, you'll learn how to develop feedback loops that involve *extended links* and that have parameters that can vary. Once you have mastered this material, it's only a matter of putting together multiple paragraphs in order to produce an insightful short story.

Appendix:
Generic Flow Templates

There are five templates that are highly useful for representing the logic of flows. These templates reappear time and again in well-constructed *STELLA* models. They are reproduced in this Appendix along with examples of each. It is well worth the time invested to understand how to construct these templates, how each works, and when it's most appropriate to use each.

- External Resource Template
- Co-flow Template
- Draining Template
- Stock-adjustment Template
- Compounding Template

External Resource

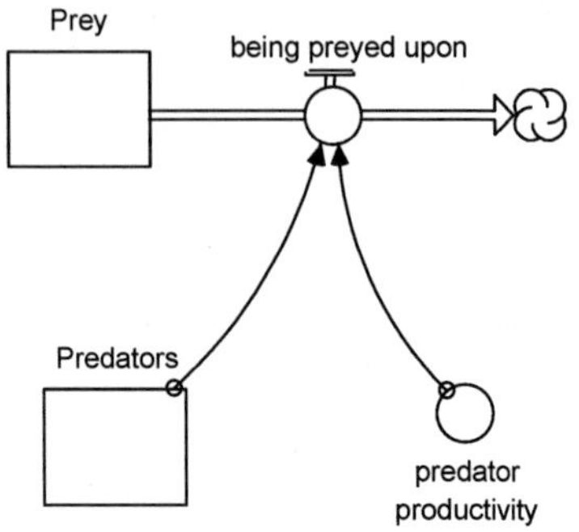

being preyed upon = Predators * predator productivity
(prey/time) (predators) (prey/predator/time)

Figure 5-7.
The External Resource Template.

Use the **external resource** template when some resource, other than the stock to which the flow is attached, provides the basis for producing the flow. Rather than the stock generating its own inflow or outflow, the flow is generated by a second stock (an "external resource"), which has an associated *productivity*.

The external resource acts as a catalyst in generating the flow (i.e., it is not consumed in the process). Below are some examples of activities well-represented by External Resource templates…

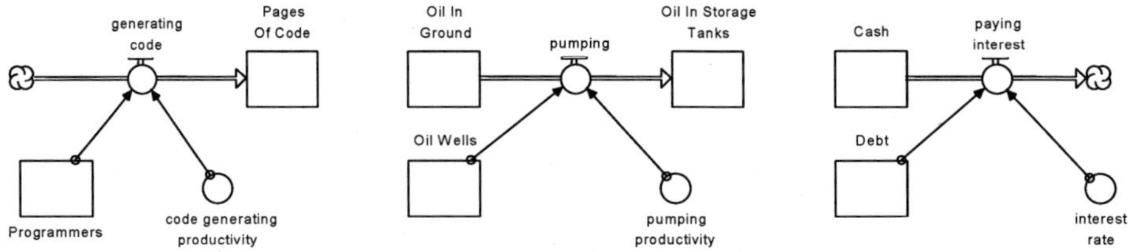

Figure 5-8.
Examples that fit the External Resource Template.

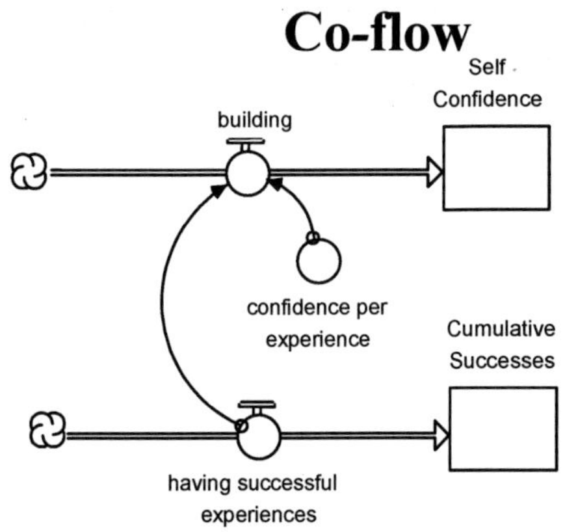

building = having successful experiences * confidence per experience
(confidence/time) (experiences/time) (confidence/experience)

Figure 5-9.
The Co-flow Template.

The term "co-flow" is an abbreviation of "coincident flow." This template is useful whenever you want to represent an activity that is driven by another activity. It is also useful when you want to track an "attribute" associated with a stock.

In a **co-flow** process, the co-flow (*building*, above) is linked to some other, primary flow (*having successful experiences).* The inputs to the co-flow process are: the "driving" flow, and a conversion coefficient (*confidence per experience*). The co-flow typically is defined as the product of the two. Some examples…

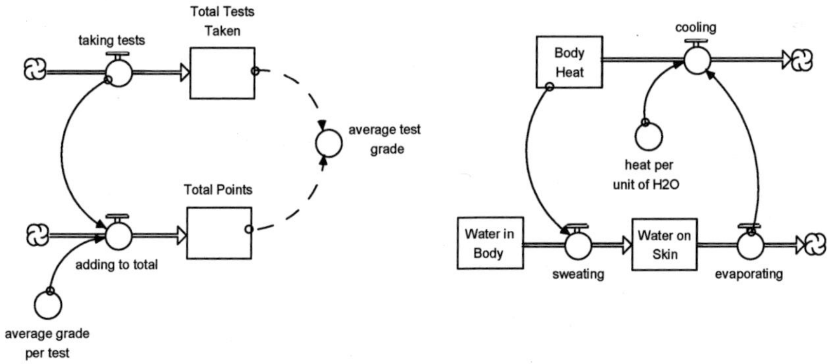

Figure 5-10.
Examples that fit the Co-flow Template.

Draining

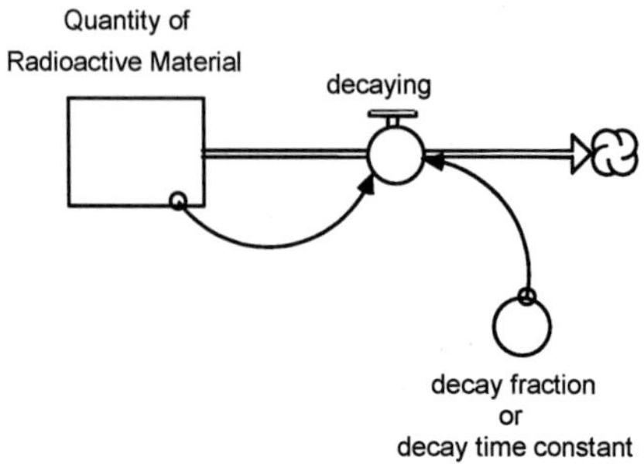

decaying = Quantity of Radioactive Material * decay fraction
(material/time) (material) (fraction/time)

decaying = Quantity of Radioactive Material / decay time constant
(material/time) (material) (time)

Figure 5-11.
The Draining Template.

Use the **draining** template whenever you want to represent a *passive decay* process. In a draining process, the flow is generated by the stock out of which it is flowing.

The flow (an *outflow* from the stock) is defined as the product of the stock and a loss fraction, or the stock divided by a "time constant." The "time constant" is the reciprocal of the decay fraction. It indicates the average length of time a unit resides in the stock, when the stock is in "steady-state." Some examples of activities well-captured by a draining template…

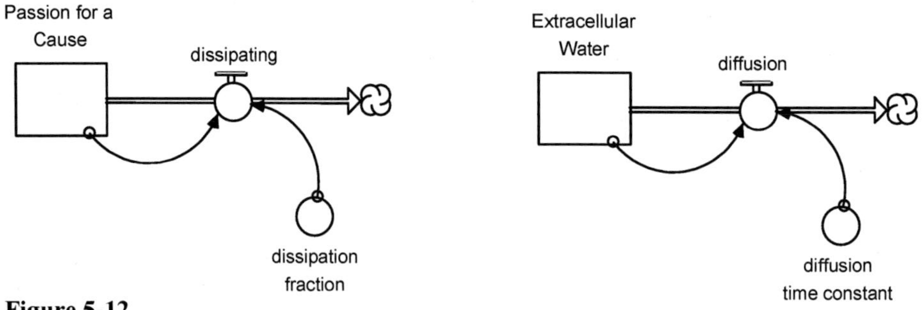

Figure 5-12.
Examples that fit the Draining Template.

Stock-adjustment

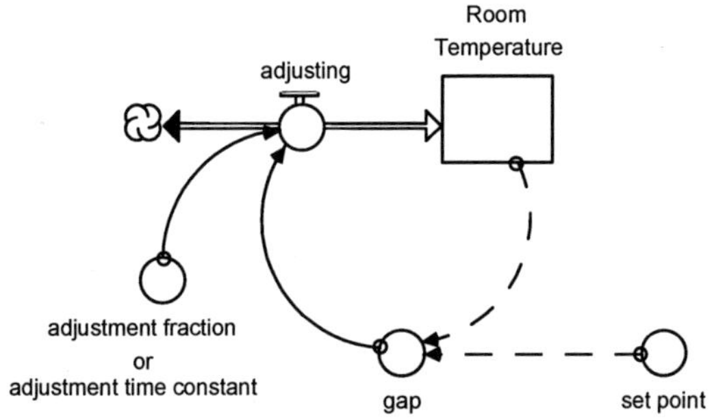

adjusting = gap * adjustment fraction
(degrees/time) (degrees) (fraction/time)

adjusting = gap / adjustment time constant
(degrees/time) (degrees) (time)

Figure 5-13.
The Stock-adjustment Template.

Use the **stock-adjustment** template to represent situations in which a Stock "adjusts to" a target value. The structure often is used to represent the way in which perceptions, opinions, and the like, are adjusted. Note that the flow is a *bi-directional*!

The flow is defined by multiplying the *difference* between the stock, *Room Temperature,* and the target *set point,* by the *adjustment fraction* (or dividing by the *adjustment time constant*). Whenever a discrepancy exists between the stock and the target, the flow will *adjust* the stock toward the target. Both the target and the adjustment fraction/time constant are usually converters, but can be stocks.

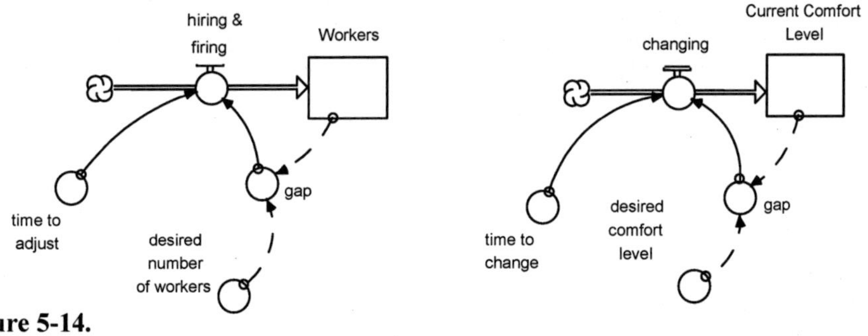

Figure 5-14.
Examples that fit the Stock-adjustment Template.

Compounding

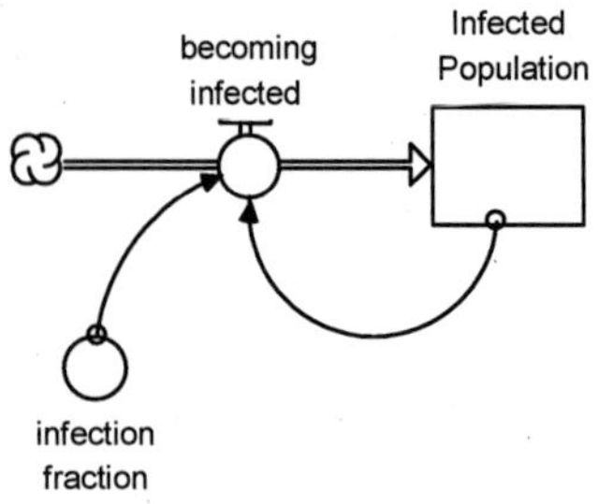

becoming infected = Infected Population * infection fraction
(people/time) (people) (people/person/time)

Figure 5-15.
The Compounding Template.

The **compounding** template is appropriate whenever you want to represent a self-reinforcing growth process. In a compounding process, the flow is generated by the stock into which it is flowing.

The inputs to the flow are the Stock (*Infected Population)* and a compounding fraction *infection fraction*. The flow into the stock (*becoming infected)* is defined as the product of the two inputs. The compounding fraction can be either a stock or a converter. Its units-of-measure are: "units/unit/time," where "units" is whatever units the stock is denominated in. The compounding fraction tells how many new units are produced by each existing unit residing within the stock, per unit of time.

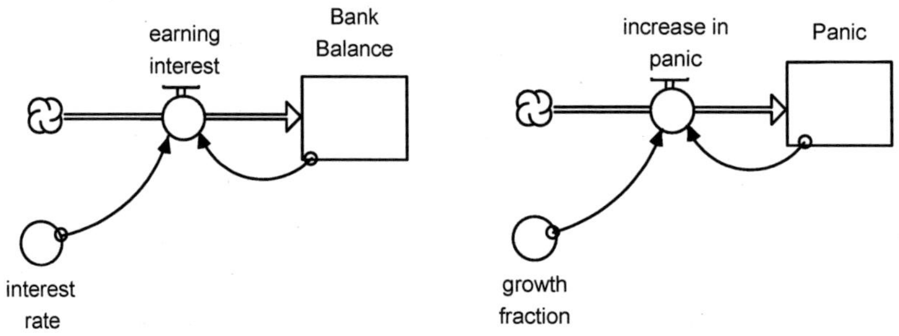

Figure 5-16.
Examples that fit the Compounding Template.

Chapter 6

Constructing "More Interesting" Paragraphs

Closed-loop & Non-linear Thinking

In Chapter 5, we looked at "simple" feedback loops. In this Chapter, we'll relax the two conditions that define loops as "simple." We'll allow the parameters associated with a feedback loop to vary, and also extend the links that constitute a feedback loop to involve more than one "sentence." As you'll see, relaxing these two constraints will enable feedback loops to generate a *much* richer variety of dynamic behavior than was possible with "simple" loops!

Allowing Parameters to Vary

Before we extend links to create feedback loops involving *multiple* sentences, let's see what behavioral richness we can engender by allowing parameters to vary within a *direct link* feedback loop structure. Figure 6-1A depicts a simple, reinforcing feedback loop (as noted in Figure 6-1 with an "R"). Left to its own doing, as we saw in Chapter 5, this loop will cause *Infected Population* to grow exponentially, forever.

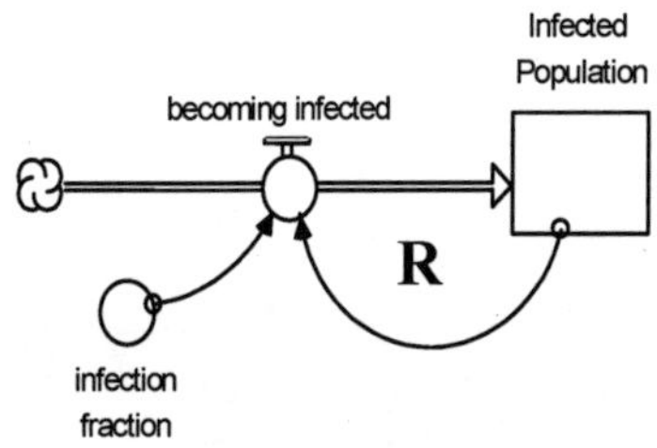

A. ***A Simple Reinforcing Loop***

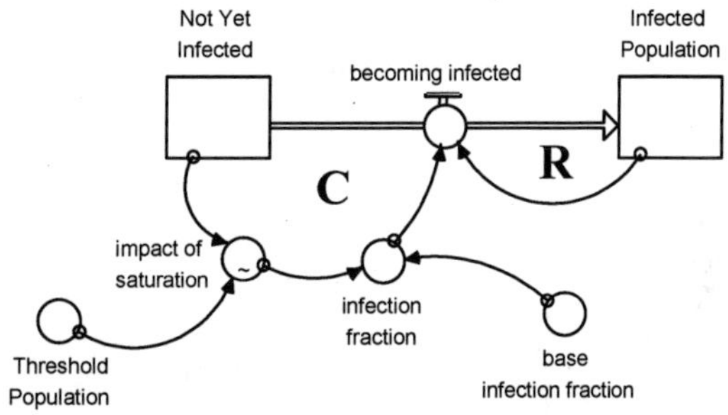

B. ***Adding a Counteracting Loop***

Figure 6-1.
From Constant, to Variable-parameter, Feedback Loops.

Figure 6-1B adds a counteracting loop to the picture(as noted in Figure 6-1 with a "C". The coupling point between the reinforcing and counteracting loop occurs in the variable, *infection fraction.* Rather than remaining constant, it is now impacted by (i.e., multiplied by) a "saturation effect," which depends on the size of the *Not Yet*

Infected population. The way in which this dependency is captured illustrates one of the most powerful features in the *STELLA* software. But before examining this feature, I want to be sure you "see" the counteracting loop (designated in Figure 6-1 with a "C").

The loop works as follows: after the *Not Yet Infected* population falls below some threshold level, the *impact of saturation* begins to depress the *infection fraction* which, in turn, slows the rate of *becoming infected*. As the *Not Yet Infected* population continues decreasing, the impact becomes more and more depressive—ultimately shutting off the *becoming infected* flow completely. This relationship reflects the fact that with fewer and fewer people who are not-yet-infected, it becomes more and more difficult to find someone who is susceptible to infection (i.e., not immune to, or protected from, the disease)! The saturation impact thus puts the breaks on the growth of the *Infected Population*. The counteracting loop "cools off" the run-away reinforcing loop. What pattern will *Infected Population* trace as a result of this interaction between a counteracting and reinforcing loop?

If you were thinking "S-shaped growth," you were correct—as Figure 6-2 indicates.

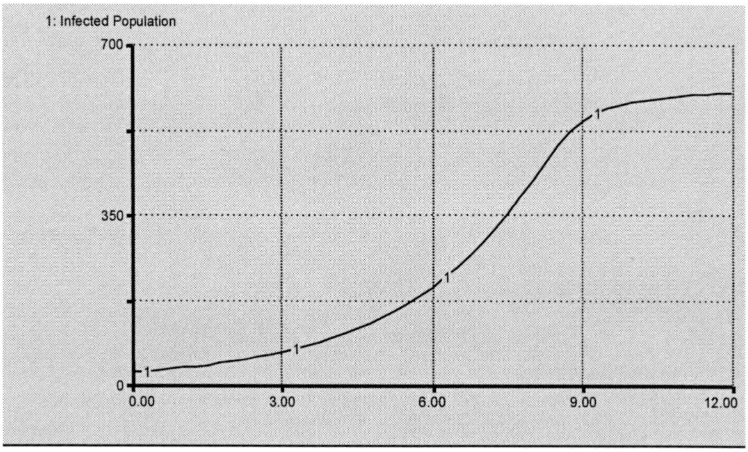

Figure 6-2.
The Behavior of the Coupled Counteracting & Reinforcing Loop.

If you weren't thinking that way, here's why you should have been… At the outset of the simulation, when the population of people left to infect is large, *infection fraction* is at its highest value. It remains constant at this value for a while because the *impact of saturation* has yet to "kick in" (i.e., the counteracting loop is exerting a neutral impact). This means the reinforcing loop is operating as if it's in isolation, compounding at a *constant* percentage rate. We therefore should expect *exponential growth* of the infected population for some period of time. And that is exactly what occurs early in the simulation. If you look at the trajectory traced by *Infected Population* up to about year 7, it is exponential. After that, it grows progressively more slowly (the top of the "S") until, by the end of the simulation, it has pretty much ceased growing altogether.

So, you might be wondering… How does that innocent-looking little *infection fraction* pull off all this magic? How is it able to exert more and more influence as the *Not Yet Infected* population decreases? It works its magic through a very important and powerful feature in the *STELLA* software. That feature is called the "graphical function."

The Graphical Function

Take a quick peek back at Figure 6-1B. If you look closely at the *infection fraction*, you will see a sign (albeit a subtle one) of its "loss of innocence" (at one time, it was just a lowly constant). Notice the little "~" on its face? A "~" designates it as a "graphical function." Graphical functions express a relationship between an input variable and an output variable. Specifically, they indicate how an output variable will change as the associated input variable changes. Importantly, they express the bi-variate relationship not by resorting to mathematics, but rather by making use of a sketchpad with a grid on it. Basically, you *draw* the relationship you envision.

Graphical functions thus enable non-mathematically-inclined people to express relationships heretofore largely limited to the domain of the mathematician. Of course, mathematicians also are welcome to use the graphical function, and many do so quite effectively. If you were to double-click on the variable *impact of saturation*, you'd see something like what appears in Figure 6-3.

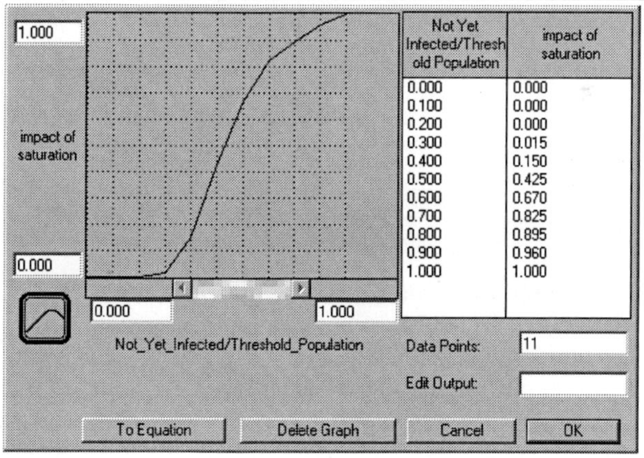

Figure 6-3.
A Graphical Function Relationship.

Graphical functions are *not* graphs of model output over time. Instead, they are used to represent "structural relationships" within the model. They show how a given variable changes as a consequence of movements in some other variable.

In this case, the *impact of saturation* is the variable being determined. The variable determining it is the ratio of the *Not Yet Infected* to the *Threshold Population.* The value the "impact" takes on will be determined by the value taken on by this ratio. When the *Not Yet Infected* population is greater than or equal to the *Threshold Population* (i.e., the ratio is greater than 1.0), the "impact" will take on a value of 1.0. Graphical functions *retain* their end-point values (i.e., they *do not* extrapolate curves beyond what is drawn). As a look at Figure 6-3 indicates, when the ratio of the *Not Yet Infected* population to the threshold value (which is a constant) falls below 1.0, the "impact" becomes progressively more depressing (remember, the "impact" is being multiplied, so a smaller value means *more* of a depressing impact). When the ratio falls to 0.2 (i.e., the *Not Yet Infected* population is 20% of the *Threshold Population*), the growth in the number of people infected is zero because the impact multiplier will take on a value of zero at that point.

As this example nicely illustrates, graphical functions enable feedback loops to change in strength over the course of a simulation. Such "changes in strength" are called "shifts in feedback loop dominance." In the absence of such shifts, the behavioral repertoire of feedback loops is pretty limited. Remember back to Chapter 5. Whenever a reinforcing and counteracting loop were both in action—one controlling the inflow to, the other the outflow from, a stock—either the stock's magnitude grew exponentially (forever), underwent exponential decay (forever), or remained unchanged (forever). That's because when compounding and draining fractions are held constant, the reinforcing loop dominates, the counteracting loop dominates, or *neither* loop dominates (because they are exactly equal in strength). And, once the initial dominance situation is created by the choice of parameter values, it then *persists* forever—because the parameter values cannot change!

Once we allow one or both parameters that determine the strength of the respective loops to vary, loop dominance can *shift*! For example, in the preceding illustration, the *reinforcing* loop is initially dominant. During its reign, the *Infected Population* grows exponentially. Then, as the population of *Not Yet Infected* falls below a threshold value, the *counteracting* loop progressively grows in strength. And, as it does, it increasingly neutralizes the *reinforcing* loop—shifting the pattern of exponential growth to a "homing in" pattern, characteristic of *counteracting* feedback loop-dominated systems.

Shifts in feedback loop dominance are one of the things that cause systems to generate "surprises." Such shifts are responsible for the "nonlinear responses" (discussed in Chapter 1) in which large pushes sometimes yield barely discernible reactions, while small tickles can unleash avalanches!

Shifts in feedback loop dominance are caused by variation in the associated parameter values (i.e., the "productivity terms") associated with the loops. In *STELLA* models, such variation is most often implemented by using a graphical function. It also is possible to vary these parameters "discretely" by using IF-THEN-ELSE type logic. Doing so, in most cases, is a violation of "10,000 Meter" Thinking. As such, I'll not treat discrete, or all-or-nothing, variation here. The *Help Files* associated with the software provide detail on how to construct such expressions. But again, for the most part, if you are embracing a Systems Thinking perspective, the graphical function will almost always be your weapon of choice for engendering shifts in feedback loop dominance.

Graphical functions are thus very important little devices. Formulating them is somewhat an art, but mostly a science. An Appendix to this Chapter conveys that science. It would be a good idea to spend some time making sure you understand the information in the Appendix—both the mechanical, and the conceptual, aspects!

Extending Links to Create "Multiple Sentence" Feedback Loops

You've now seen how relaxing the assumption of constancy, with respect to the parameters that determine the strength of a feedback loop, can enrich the repertoire of dynamics a system can exhibit. The next bit of relaxation will be with respect to the "extent" of the feedback linkages themselves. All of the feedback loops we examined in Chapter 5, and up to this point in this Chapter, have included only *one* sentence (albeit in some cases a compound sentence). We're now ready to see what can happen when we extend the links constituting a feedback loop to include *more than one* sentence.

Figure 6-4 illustrates a two-sentence feedback loop structure. The loop is *counteracting* in nature. Its purpose is to maintain *Inventory* at a target level. The "strategy" for doing so is to adjust *Labor* upward or downward so as to regulate the volume of *producing*. Here's how the feedback loop works…

Initially, the *selling* volume is constant. *Labor* is at a level that causes the *producing* volume (note that *productivity* is *constant*) to just exactly equal the *selling* volume. As a result, *Inventory* remains constant at its "target" level (the target is "buried" inside the *hiring & firing decision* DPD). As long as *Inventory* remains at target levels, the *hiring and firing* volume will remain at zero. And, as long as the *hiring\firing* volume remains at zero, *Labor* will remain constant. The system is in steady-state.

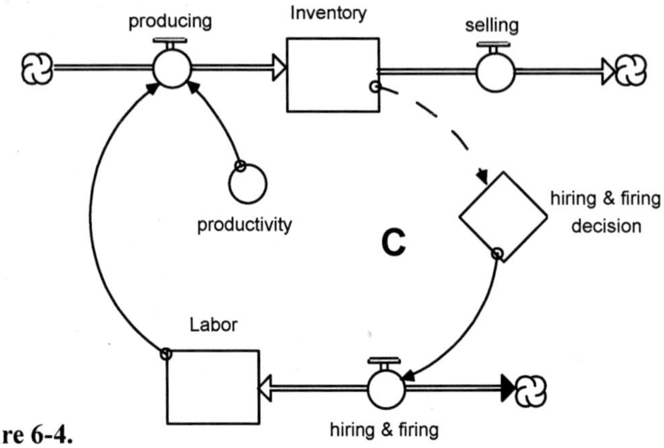

Figure 6-4.
A Two-sentence, Counteracting Feedback Loop.

But you know how Systems Thinkers *hate* systems that remain at rest. They want to see dynamics! To coax this system into strutting its stuff, we'll step-up the formerly-constant *selling* flow to a higher, constant volume. Mentally simulate what you think will unfold in response to this disturbance.

Did you guess the pattern that you see in Figure 6-5? If so, bravo! Most people don't.

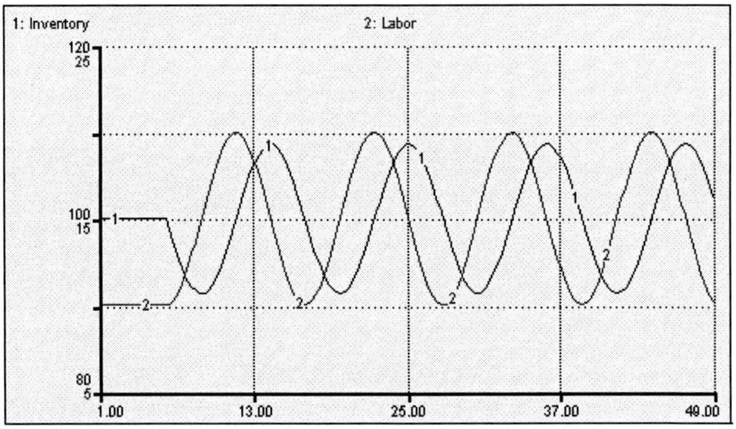

Figure 6-5.
The Response of the Inventory/Labor System.

I'll offer a brief "anatomical" explanation here. But first, recognize that this simple-looking little structure generates some pretty wild and wooly behavior! Linking sentences via feedback loops—even with *constant* parameters—really expands the associated behavioral repertoire!

In this case, what's going on is this…as soon as the step-increase in *selling* (the *outflow* from *Inventory*) occurs, *Inventory* starts falling because the *producing* flow volume (the *inflow* to Inventory) remains at its previous steady-state value—which is less than the now stepped-up *selling* volume. As *Inventory* dips below target levels, hiring "kicks in." The resulting increase in *Labor* drives up the *producing* volume. However, until the *producing* volume increases to equal the *selling* volume, *Inventory* will continue to fall. Make sure this much makes sense before continuing.

Okay, so what happens to *Inventory* at the point when hiring has caused an increased stock of *Labor* to drive up the *producing* volume to the point where it now, once again, just equals the *selling* volume?

If you said, *Inventory* ceases falling…Bene! However, notice something that's critical to understanding these dynamics. At the point where *Inventory* has ceased falling, it is as far away as it's ever going to be from its target level! Do you see this?

And so, when the *producing* volume has increased to the point where it is (as it was initially) equal to the *selling* volume—a necessity for steady-state to be re-achieved—*Inventory* is as far as it's ever going to be from its steady-state level! We thus now have a system that is very seriously out of whack! The flows associated with a given stock are equal at exactly the point where the stock is as far as its ever going to be from its target level! This is precisely the condition that must prevail in order for what's called a "sustained oscillation" to occur.

Let's continue a bit more to ensure you understand what's going on. When *Inventory* is as far below its target level as it's going to be (i.e., the negative discrepancy between the two is at a maximum), the rate of hiring will reach a maximum. This means that the stock of *Labor* is expanding at its maximum rate, and hence that *producing* likewise will be *increasing* at its maximum rate—and this is occurring right at the point where *producing* is just equal to *selling*. So, as the stock of *Labor* continues to expand, the *producing* volume will follow suit, soaring right on past the *selling* volume. And, as it does, *Inventory* will begin to re-build (the inflow to the bathtub will now exceed the outflow), and the rate of hiring will hence slow.

However, as long as *Inventory* levels remain below target levels, hiring will continue and hence the *producing* volume will continue to increase beyond the *selling* volume. At some point, *Inventory* levels will have re-built back up to exactly target levels—another of the conditions necessary for the system to re-gain its steady-state. But, at this point, can the system come to rest?

Uh-uh. Because at this point, the *producing* volume—though it will now cease increasing—will stand as high as it's ever going to be *above* the *selling* volume. That means there's way too much *Labor* on board. So, while *Inventory* will have re-achieved its steady-state level, *Labor* will be as far as it's ever going to be above its steady-state level. Do you see the problem? It's called "out of phase" goal-seeking! And given this feedback structure arrangement, goal-seeking can never get back "in phase."

Bottom line: Although this system is being regulated by a *counteracting* loop, that loop is not capable of returning the system to rest. It will try. It will goal-seek its heart out! But because of the nature of the counteracting relationship, this system will continue to oscillate for eternity (or for as long as your laptop battery lasts, if you're on an airplane).

Combining Variable Parameters *and* Extended Links

So how can we gain some measure of control over this system? Counteracting loops are always a better bet for increasing control than reinforcing loops, so let's add a *second* counteracting loop to this system. We'll do so by allowing one of the previously constant parameters to become variable.

The particular parameter we'll allow to vary is *productivity*. *Productivity*, in effect, determines the *strength* of the connection between *Labor* and the *producing* flow (which is to say, the strength of the counteracting feedback loop). That is, the larger the value *productivity* takes on, the smaller the amount of *Labor* that will be needed in order to elevate the *producing* volume by a given amount (because each *unit* of *Labor* will contribute a larger amount to the *producing* volume). Conversely, a smaller value for *productivity* will weaken the counteracting loop because it would mean that more *Labor* would need to be brought on to boost *producing* by any given amount.

Suppose we were able to strengthen the counteracting loop both by boosting *productivity* whenever the *producing* volume needed to increase, and by lowering *productivity* whenever that volume needed to be cut back. Such variation occurs naturally in most work situations. Swollen work backlogs tend to inspire focus, buckling down, and getting the job done (i.e., *productivity* rises). Lean backlogs enable people to drink more coffee and share more water cooler conversation (*productivity* falls).

To implement such a variation capability, we'll rely upon our old pal the graphical function. The resulting new "structure" looks like what you see in Figure 6-6.

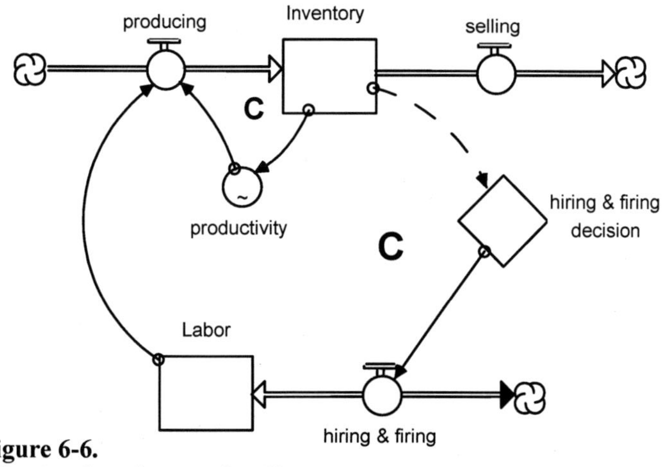

Figure 6-6.
Allowing "productivity" to Vary.

Notice that by linking *Inventory* to *productivity*, we've added a second *counteracting* feedback loop. This second loop works in concert with the first one, which is to say, amplifies its strength! This second loop carries some of the burden of causing the volume of the *producing* flow to increase (and decrease) as *Inventory* levels rise and fall with respect to target. For example, rather than having to crank up *producing* solely by bringing on additional labor, some of the increase in *producing* can be delivered via elevating *productivity* levels.

So, what effect do you think adding this second counteracting loop will have on the system's behavior? Will it heighten or dampen the instability the system is exhibiting? And why?

Such questions confound intuition. One of the significant contributions of the *STELLA* software is that it provides a check on intuition, while also providing a vehicle for building an understanding of "why." In the process, your capacity for intuiting dynamics will be developed—as will your ability to articulate the associated "how comes." Figure 6-7 shows what happens to the system's behavior when the second counteracting loop is added.

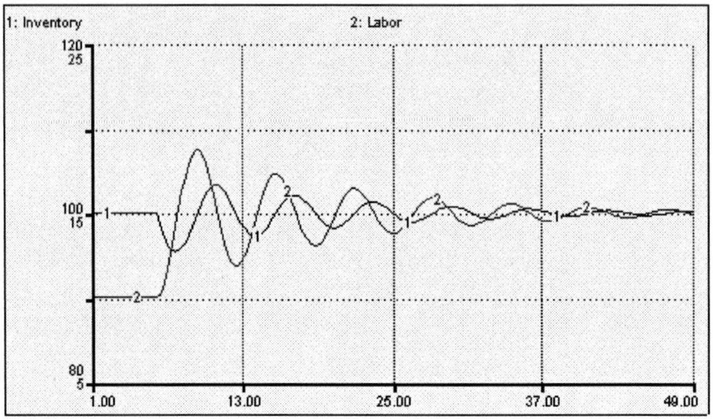

Figure 6-7.
From Sustained to Dampened Oscillation.

As the Figure shows, activating the new loop dampens the oscillation. A simple explanation is that instead of *Labor* having to continue to expand to the point where *producing* is at its maximum above *selling*, *producing* can now reach its maximum above *selling* before *Labor* reaches its maximum (because *productivity* is *also* boosting *producing*!). This means that not as much *Labor* will be hired on the upswing, and hence not as much will need to be shed on the downswing. That, in turn, means even fewer people are hired on the *next* upswing, and thus even fewer still are shed on the subsequent downswing; and so on. The system is thus able to progressively settle back down into a steady-state (barring further externally-produced disturbances).

Feedback Loops, In Summary

A feedback loop is an ingenious and incredibly powerful "structure." Feedback loops abound in physical, technological, natural, and social systems. They enable these systems to maintain internal balances, and also to grow. They guide evolutionary adaptation, and preside over catastrophic collapses. Feedback loops *self-generate* all manner of dynamic behavior. Excite one and you will set in motion an ongoing dynamic, not a one-time response. The pattern that dynamic will trace depends on the relative strengths of the various feedback loops that make up the system, and how those strengths wax and wane over time. The graphical function in the *STELLA* software, by serving as a coupling point between loops, is often the vehicle for enabling such waxing and waning to unfold.

You are now well-prepared (after studying the Appendix to this Chapter) to begin capturing in your *STELLA* models the feedback loops that exist in the realities you are seeking to represent. You will see lots more examples of feedback loops throughout the remainder of this Guide and in the sample models that accompany the software. Capturing the feedback loop structure of a system, in an operational way, is the essence of the difference between building models with tools like spreadsheets versus using the *STELLA* software. It's an important difference!

What's Next

In the next chapter, several examples of generic feedback loop structures are provided. You will find these little "infrastructures" to be very useful as building blocks for populating the "short stories" you write with the *STELLA* software.

Appendix: Formulating Graphical Functions

This Appendix describes two principles to keep in mind when formulating graphical functions, and then goes on to provide a "cookbook" of steps to follow in formulating them.

Principle 1. *The Ceteris Paribus Principle*

Graphical functions are used to capture a relationship that you hypothesize to exist between two, and only two, variables whose interaction you are thinking about against a "ceteris paribus" (all other things held constant) backdrop. When you sketch into the graphical function the curve you feel captures the relationship you are seeking to represent, the slope of that curve should (in general) not change direction! If it *does*, think hard about whether you are not implicitly including the impact of one or more other variables in your formulation of the relationship. Let's take an example to clarify the point.

Schedule pressure is often brought to bear on workers when a project falls behind schedule. The idea is that such pressure can cause people to increase their focus on the task at hand, and hence increase their *productivity*—speeding the project forward, and hopefully putting it back on schedule. A description of the relationship between levels of *schedule pressure* and resulting levels of *productivity* usually goes something like this…

In the complete absence of *schedule pressure*, *productivity* is less than it could be because people will not focus well without feeling some pressure from a deadline. As *schedule pressure* rises up from zero, *productivity* increases for a while. But, beyond a certain point, *schedule pressure* becomes dysfunctional because it weighs too heavily on workers. Implementing the preceding logic into a graphical function would yield something like what you see in Figure 6-8.

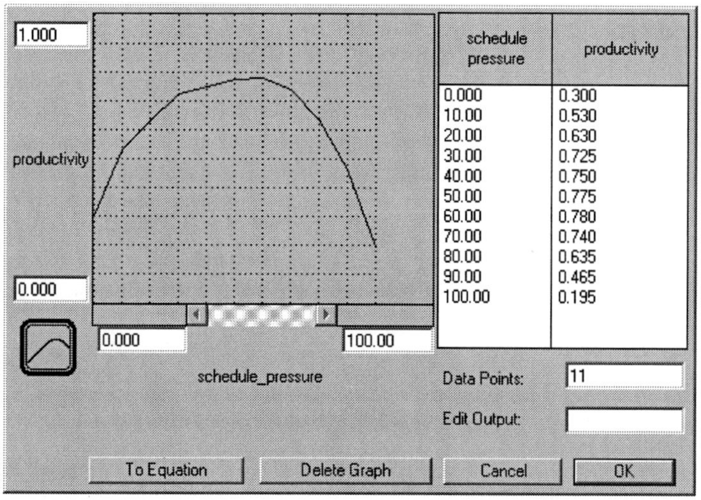

Figure 6-8.
A Schedule Pressure/Productivity Graphical Function.

Clearly this is a curve whose slope changes direction. Let's think more carefully about the assumptions that underlie it. How does schedule pressure *directly* impact workers? Not, what actions does schedule pressure cause workers to take, but how does it directly impact them? It is this latter question that you want to ask when formulating a graphical function.

Schedule pressure usually takes the form of a project manager reminding workers of impending deadlines at a higher frequency than normal, and also with a greater sense of urgency. This may, in fact, cause workers to take certain actions as a result. They may, for example, work longer hours, take fewer breaks, focus more on the tasks at hand, and so forth. Once they start doing such things, they don't stop doing them as schedule pressure mounts because of the schedule pressure, itself! They may stop doing them because they get fried, or miss time with their families, or for other reasons—but not because schedule pressure is increasing!

The *direct* impact of schedule pressure on productivity is therefore probably to increase it—though the impact certainly saturates. Schedule pressure can't have a positive impact, and then at some "magic point" all of a sudden reverse the direction of its impact! The perceived change in direction of impact is, in fact, due to unconsciously introducing other things (like burnout, waning motivation, etc.) into the thinking process—and they don't belong in the *same* graphical function!

If you do not carefully screen out such "other influences" on an output variable when formulating a graphical function relationship, your models can yield misleading conclusions when you simulate them. For example, in the preceding illustration, what if workers *already* were experiencing some level of burnout from, say, a previous project. Now, assume they just begin to fall behind on the current project (i.e., zero schedule pressure had been applied). The graphical function relationship described in Figure 6-8 would suggest that you could *increase* worker productivity by applying some schedule pressure. However, clearly that would *not* be the outcome if workers already were frayed at the edges!

Be certain the "thought experiments" you conduct in formulating graphical functions involve two, and only two, variables! Screen out "other influences" on the output variable.

Principle 2.
Think Out of the Historical Box

The second important principle to follow when formulating graphical function relationships is: *Be sure to estimate any relationship over its full potentially realizable operating range, and not just over a range that may have been historically-observed.* Many people feel uneasy about formulating graphical function relationships in general. This is, in part, due to the fact graphical functions often are used to capture "squishy" relationships. But even when the relationships are more tangible—such as, say, the impact of price on demand—people often have issues with venturing outside historically-observed ranges for the relationship. For example, historically, price may only have ranged plus or minus 25% from its current value. And so, this is the extent of the range many people feel comfortable with including in the associated graphical function. "Beyond this range, we have no solid empirical data," is what we often hear.

But think about it...If you want your model to be able to shed any light on what *could* happen if price were dropped to zero, or boosted by 50%, you *must* provide *some* estimate of price elasticity over this range! Oftentimes, real insights emerge when you drive a system to operate outside its historical operating range. If your model's parameter values (especially its graphical function relationships) are not estimated over their full, potentially-realizable operating range, you forfeit the opportunity to have the model "surprise" you! Model results, in these cases, could be "crazy." But with a *STELLA* model, it's always possible to discover how those results are being brought about, and you can therefore always separate "craziness" from genuine "new insight." I have witnessed the latter on a sufficient number of occasions to feel very strongly about the importance of principle number two.

Cookbook Guidelines for Formulating Graphical Functions

Let's now look at some general guidelines to follow when formulating graphical functions. The guidelines that follow are arrayed in a progression of steps. Following these steps, in the order presented, will generally enable you to formulate reasonable graphical function relationships, whether you are doing so alone, or via group discussion.

Step 1. Apply Ceteris Paribus Principle

Think only of the relationship between the input variable and the output variable, holding all other variables impacting the output variable constant.

Step 2. "Normalize" the Input Variable

It's not always necessary to "normalize" the input variable, but it is so often useful, I list it as a step.

"Normalizing" is accomplished by dividing the input variable by some *appropriate* quantity. Percentage variables (like market share) and index-number variables (e.g., variables that are scaled 0-100, like motivation, self-esteem, and burnout) need not be normalized.

Normalizing has a couple of important benefits. First, it makes movements in an input variable easier to think about because a normalized range usually extends from 0 to 1, or 0 to 2, rather than from 0 to 10000, or 500 to 5000. When a range is 0 to 2, it's much easier to think about changes in the input variable in *percentage terms*. That is, if the input variable increases from 1 to 1.25, that's immediately recognizable as a 25% increase. If an input variable's range is, say, 0 to 10,000, it's difficult to know at a glance how much of a percentage increase a move from, say, 570 to 730 constitutes.

A second benefit from normalizing is that it makes the relationship "scale independent." If you used *absolute* ranges for input variables, you would have to re-calibrate your graphical functions any time those absolute ranges changed. By normalizing, you convert to a *relative* scale. For example, if Deer Population were your input variable, the question might shift from "If Deer Population falls to 4,328 …" to "If Deer Population falls to 50% of its beginning of year 2001 value.

Choosing an appropriate "normalizing" variable often takes a little thought. Sometimes, simply dividing the input variable by its starting value (i.e., its value at the outset of the simulation) works quite well. Other times, dividing through by a variable that has different units-of-measure works better—such as deer per wolf, or grams per milliliter.

Step 3. Establish Ranges for the Input and Output Variables (apply Principle 2)

Be sure to establish ranges that permit full possible movement of both input and output variables, not just movement that has been historically-observed. Remember that graphical functions do not extrapolate outside their defined ranges. Instead, they retain the first and last output values that have been assigned. If you do not extend these ranges to cover full possible movement, you are limiting the space for producing model-generated insight!

Step 4. Determine the Direction and Nature of the Slope

Remember that if the slope of the curve you sketch *changes* direction, you are probably including more than one input variable in your thinking. If the slope of your graphical function changes direction, think very carefully about your assumptions, and then run the thinking by a few other people.

By "nature" of the slope, I mean...Does the curve saturate? Is it linear? S-shaped? And so forth. Make explicit a behavioral argument to support your choice of a curve, and include it in the Document cache of the graphical function equation dialog so that others can understand your rationale.

Step 5. Identify Extreme Points and, if Appropriate, a "1,1" Point

Begin with the low-end x-point (input value), and establish the associated y-point (output value). Then, do the same for the high-end x- and y-points. In some cases, particularly if you are using "impact of..." variables, you will also be able to establish a so-called "normal point" or "1,1" point. When an "impact of..." variable (usually a "multiplier") takes on a value of 1.0, it means it is exerting a *neutral* impact. A normalized *input* variable usually takes on a value of 1 in the initial condition, when the system is in steady-state.

Step 6. Sketch a Smooth Curve Through the Established Points

Whether you have only two extreme points, or those two points plus a "normal point," trace some sort of *smooth* curve through the points. If you have some "magic points" in your curve (some points at which sharp discontinuities occur, or the slope changes direction), think very carefully to be sure you can justify what you've drawn.

Chapter 7

Short Story Themes

Generic Infrastructures

This Chapter concludes the progression begun in Chapter 2 by presenting a few "short story themes"—a set of five generic *infrastructures* that can serve as nuclei for models that you construct. Each is built from a combination of paragraphs. Each gives rise to its own interesting dynamic behavior. We make use of these infrastructures in many of the models we construct. We encourage you to do the same wherever appropriate.

The infrastructures are arrayed in template form, and are contained as *STELLA* models in the *Intro to Systems Thinking* folder. In that folder, you'll find templates for the following infrastructures:

- Overshoot and Collapse
- Slippery/Sticky Perceptions
- Main Chain
- Attribute Tracking
- Relative Attractiveness

For each infrastructure, I'll provide…

- Suggestions for when to use it
- A description of the structure
- An explanation of the dynamic behavior
- Some variations on the generic theme

Overshoot and Collapse

When to Use

Many physical, biological, and social systems do not make a smooth transition from growth to steady-state. Instead, they grow rapidly, peak, and then collapse to a lower steady-state value—which, in some cases, is extinction.

Description of Structure

Figure 7-1 depicts a simple overshoot and collapse structure.

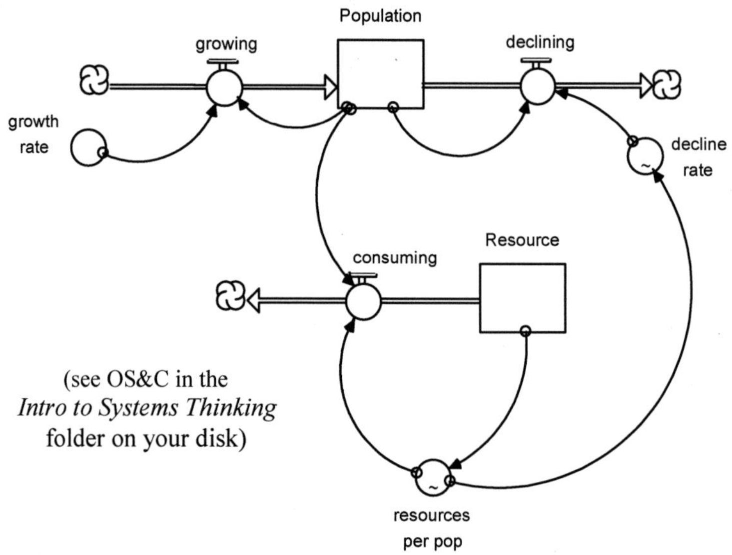

(see OS&C in the *Intro to Systems Thinking* folder on your disk)

Figure 7-1.
A Simple Overshoot & Collapse Structure.

A population consumes a non-renewable resource. The resulting scarcity of the resource then drives a collapse of the population. No recovery is possible because the resource is not renewable.

Description of Behavior

The overshoot and collapse pattern of behavior is shown in Figure 7-2. As the Figure indicates, the growth phase of the pattern looks much like S-shaped growth. Growth expands rapidly at first. At the outset, the *Resource* is abundant, so the compounding process dominates the behavior. As the *Resource* is drawn down, the death rate loop gains strength. Growth slows as the *Population* approaches its maximum value. But the *Population* cannot be sustained because the *Resource* will continue to be depleted as long as there is any amount of

Population consuming it. As the *Resource* continues its inexorable decline, the *decline rate* races by the *growth rate*. Thus, the outflow from *Population* becomes greater than the inflow, and it will always remain so.

Three curves are shown in Figure 7-2. The first is a Base Case. The second and third were produced by doubling, then tripling, the starting amount of *Resource*. Notice that doing so does not come anywhere close to doubling or tripling the amount of time before the system collapses! That's because compounding processes generate exponential, not linear, growth! Increasing the initial amount of resource simply allows compounding to go on for a longer amount of time. It doesn't take long for an exponential to soar out of control.

Variations

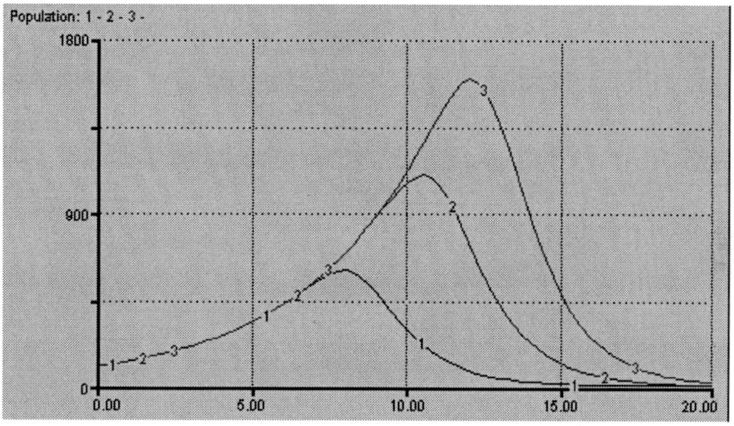

Figure 7-2.
Overshoot & Collapse with a Doubling & Tripling of Initial Resource.

Figure 7-3 provides a useful variation of the generic overshoot and collapse structure. It allows for the resource to regenerate, but the regeneration rate *declines* as the level of the resource *declines* (an example: the girdling of trees by deer). This structure can generate a rebound from the collapse under certain situations, as you'll see if you run the associated model. In order for a rebound to occur, it's essential that *resources per pop* go to zero before *Resource* reaches zero. Otherwise, once *Resource* hits zero, the system has no basis for regenerating.

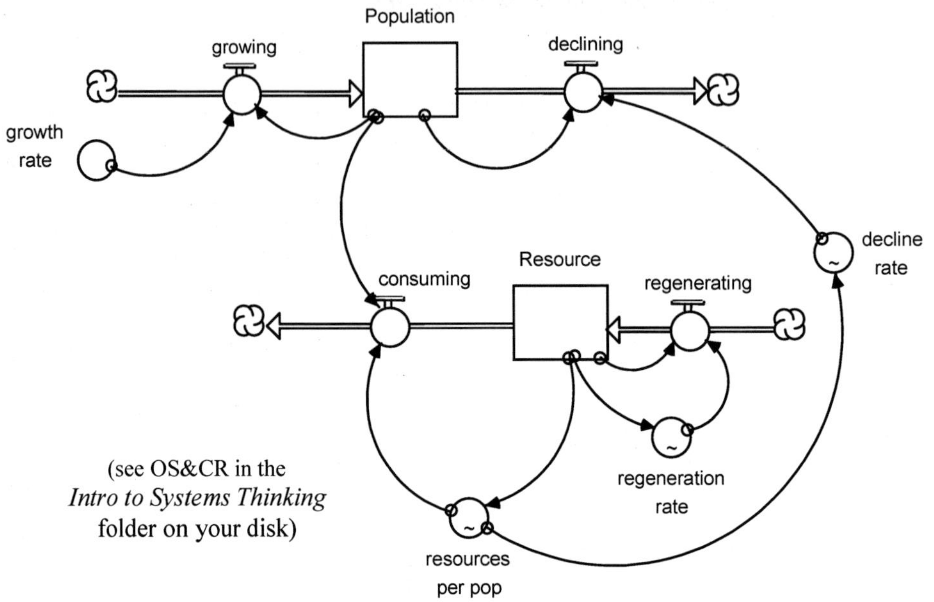

(see OS&CR in the
Intro to Systems Thinking
folder on your disk)

Figure 7-3.
Making the Resource Renewable—with a Variable Regeneration Rate.

Slippery/Sticky Perceptions

When to Use

When there is an asymmetry in the rate at which perceptions are adjusted, this structure can capture it nicely without the need for a lot of technical wizardry. An example would be when consumers become aware of a quality problem with a product. They generally adjust their perceptions of the product's quality downward very quickly. Then, when the company who produces the product fixes the quality issue, it often takes much longer for consumers to re-adjust their perceptions back upward. The same often is true for, say, the perception of a person's reputation. If the person "messes up," many will quickly lower their perception of the person's character, and it will take a lot of evidence to the contrary to rebuild the former perceptions.

Description of Structure

The Slippery/Sticky structure is depicted in Figure 7-4. The astute observer will quickly recognize it as a stock-adjustment template with one "wrinkle." The *adjustment time* is a variable, rather than a *constant*. Specifically, the *adjustment time* depends on the relationship between the *Perceived State* and the *actual state*. If the *actual state* were less than the *Perceived State*, the *adjustment time* would be small—assuming you wanted to capture a perception process that was slippery *downward*. If the *actual state* was greater than perceived, the *adjustment time* would be large—capturing a "sticky upward" adjustment.

(see SlSt in the *Intro to Systems Thinking* folder on your disk)

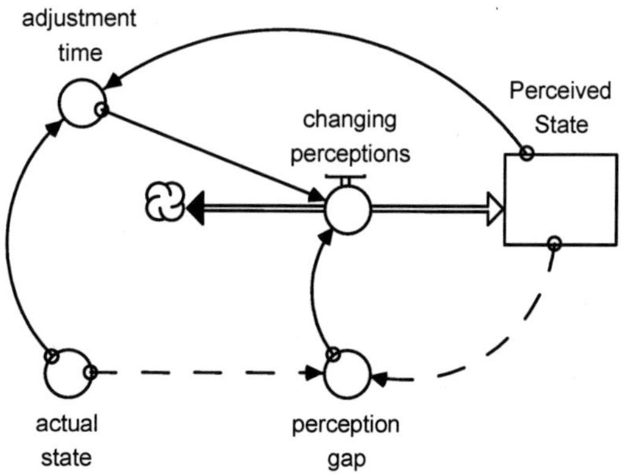

Figure 7-4.
A Sticky/Slippery Perception Process.

Description of Behavior

This behavior generated by this infrastructure is shown in Figure 7-5. What's depicted is the response, from an initial steady-state, to a 40%

step-increase and step-decrease in the *actual state*. As you can see, the two responses are *not* symmetrical. The downward adjustment is completed in a few time periods (slippery) while the upward adjustment takes nearly 50 time periods to complete (sticky). If you look at the actual model on your disk, you will notice the use of some Boolean algebra to define the *adjustment time*.

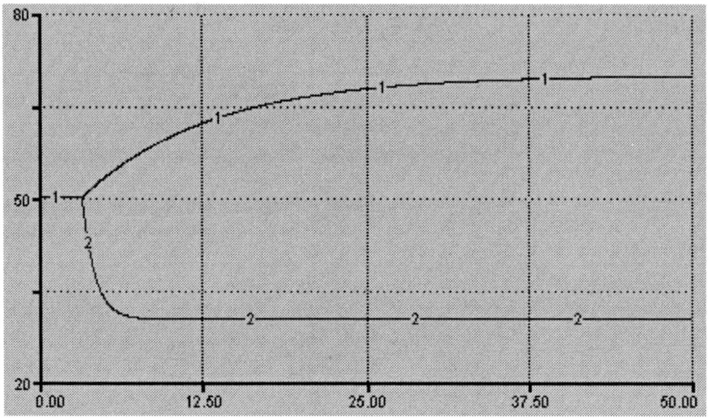

Figure 7-5.
The Slippery/Sticky Behavior Pattern.

Variations

One variant on this generic infrastructure is to allow the *adjustment time* to be represented by a graphical function, rather than defining it using Boolean algebra. This will allow the speed of adjustment to vary more continuously, rather than just being one value if actual is less than perceived, and another value otherwise.

Main Chain

When to Use

The *main chain* infrastructure, also referred to as a "spinal cord," is quite useful whenever you want to represent a sequence of stages through which stuff passes. A few of the many examples include: the sequence of phases a plant or animal goes through as it ages; the psychological states individuals pass through en route to accepting a loss or failure; the progression of steps in a human resource promotion chain; and the stages a market innovation passes through as it "matures."

Description of Structure

Figure 7-6 illustrates a common use of the main chain infrastructure: representing an aging process.

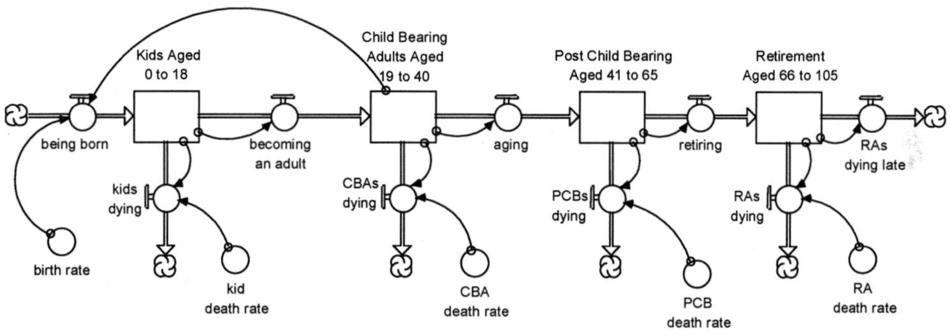

(see MC in the *Intro to Systems Thinking* folder on your disk)

Figure 7-6.
An Aging Chain.

In this configuration, the chain of reservoirs is fed at the front-end by a single flow, *being born.* Two outflows drain each reservoir. The first is a flow-through that moves stuff on to the next "phase" (age category, in this case). The second is an exit flow, which drains stuff out of the chain. Associated with each exit flow is an age-cohort-specific death rate. Note that all outflows in the chain are represented using the Draining template.

Description of Behavior

Main Chains exhibit an interesting characteristic behavior pattern. In steady-state, they will distribute their total contents among the stocks in the chain in proportion to the average residence time associated with each stock. Each stock's average residence time is determined as some blend of its flow-through and exit "time constants."

The graph in Figure 7-7 illustrates the behavior of the illustrative main chain from an initial steady-state starting point. At time 5, the death rate associated with the last stock in the chain is stepped down. This causes the average residence time associated with that stock to increase (people remain in the stock, on average, for a longer period of time). As a result, as the Figure indicates, the percentage of the total population in the last class increases to a new higher, steady-state level.

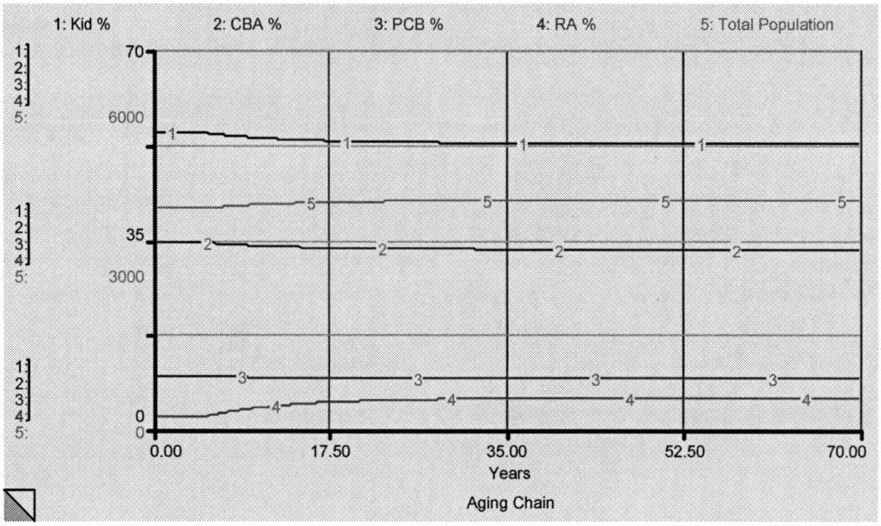

Figure 7-7.
Shifting the Steady-state Distribution of Content in a Main Chain.

Variant

In the "classic" form, the Main Chain is fed with an inflow only into the first stock in the chain. Adding inflows to any of the other stocks is one way to modify the classic structure. Another option is to replace the reservoirs with conveyors. Finally, in many situations the parameters associated with the draining processes (i.e., the draining fractions) vary—rather than remain constant.

Attribute Tracking

When to Use

The *attribute tracking* infrastructure is useful whenever you want to "track" an attribute associated with a stock. The structure creates a moving "exponential average" of the attribute. Unlike an arithmetic average calculation, which weights every number going into it equally, an exponential average gives progressively less weight to the further back in time numbers being used in the calculation.

Description of Structure

The map in Figure 7-8 shows the basic attribute tracking structure.

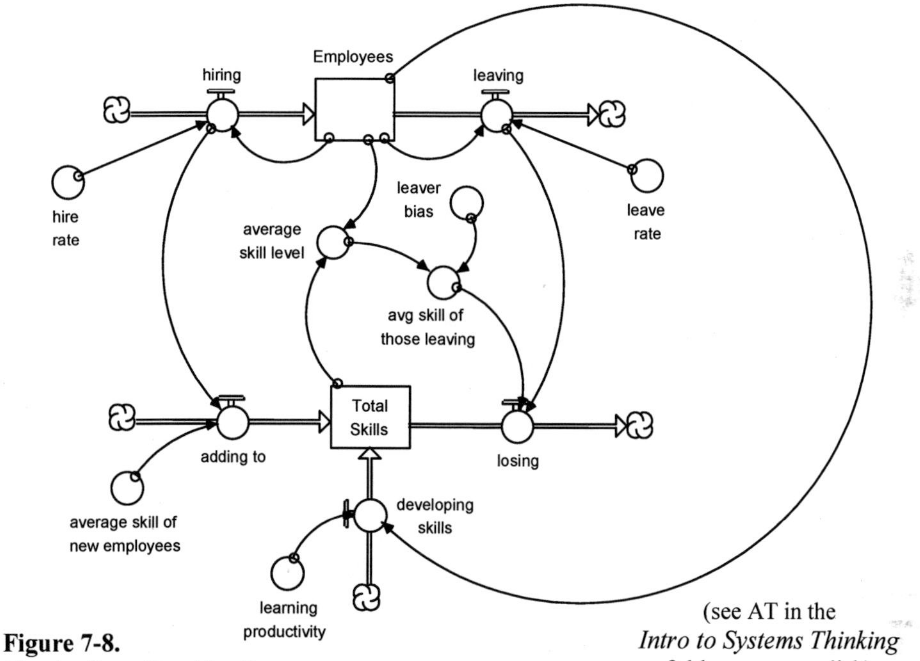

Figure 7-8.
The Attribute Tracking Structure.

(see AT in the *Intro to Systems Thinking* folder on your disk)

In this case, the attribute being tracked is the skill level of a population of employees. To calculate a moving average skill level for the overall population, the total number of employees is divided into the total amount of skills they possess to yield an average skill level per employee. Each employee who is hired carries with them an average amount of skill (a co-flow process). Each one who leaves, also through a co-flow process, carries with them an average level of skill. In this illustration, as the Figure indicates, this latter amount is related to the current average skill level of the population. The relationship is expressed by multiplying the average by a *leaver bias*. If this bias is greater than 1.0, leavers take something greater than the current average. If the bias equals 1.0, leavers take the average. And, if the bias is less than 1.0, leavers take less than the average when they depart. There is one other inflow to the stock of *Total Skills*, and that is *developing skills*. This flow occurs independently of the flows of employees. The *developing skills* flow is formulated as an External Resource process, though this need be not always the case.

Description of Behavior

The infrastructure is initialized in steady-state. In steady-state, *hiring* equals *leaving*. The *leaver bias* is zero, meaning that those leaving are taking with them the existing employees' average level of skills. However, new employees come into the organization with a lower skill level than the existing employee population. Thus, in order to remain in steady-state, the system must offset this difference through the *developing skills* inflow.

At time period 3, the leaving bias steps up to 20—meaning leavers now begin taking 20% more than the average skill level of existing employees. As Figure 7-9 shows, the result is a slow decay of the average skill level of the existing population down to a new, lower steady-state value.

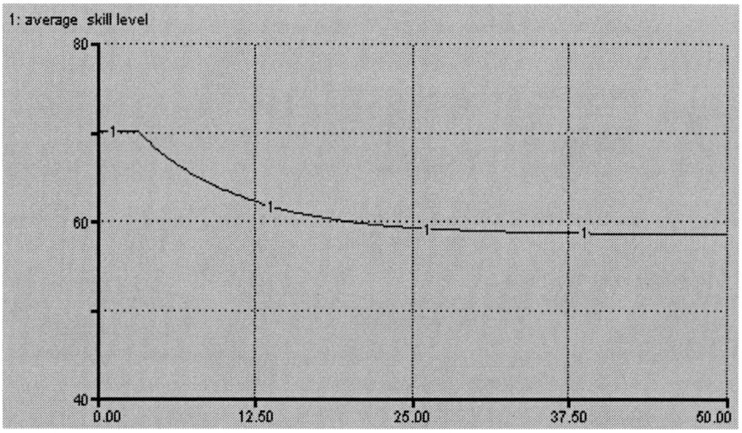

Figure 7-9.
The Response of the Attribute Tracking Structure to a Step-increase in Leaving Bias.

Variations

The most popular variations on this structure are achieved by allowing each of the constant parameters associated with the structure to vary. Often the variations are "driven" by the average level of the attribute. So, for example, in this illustration, the *leave rate*, *leaver bias*, *learning productivity*, and the *avg skill of new employees*, could all be represented as graphical functions of *average skill level*.

Relative Attractiveness

When to Use

The *relative attractiveness* infrastructure is useful whenever you want to generate an index of attractiveness that is comprised of a set of attractiveness components. The index consists of a weighted average of these components.

Description of Structure

Figure 7-10 illustrates the relative attractiveness structure. The context for the illustration is a population center (e.g., town, city, region, country). Only three attractiveness "factors" are illustrated, but the structure is easily extensible to as many factors as you would want to consider. Note that the three weighting factors are attached to sliders. If you open the model, you will find that the sliders are "chained," meaning that no more than 100% of the weight can be distributed.

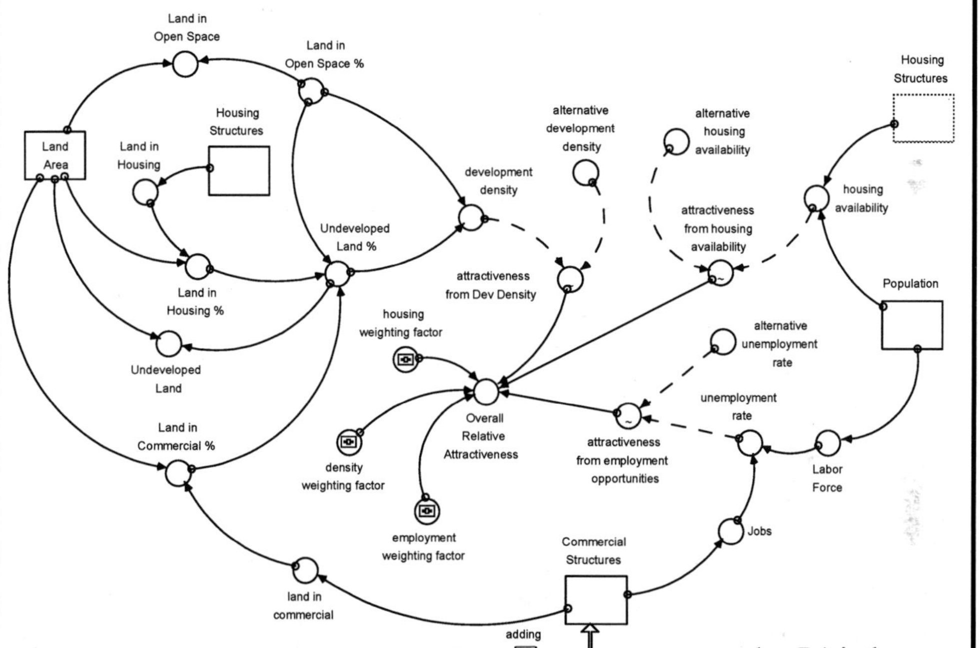

(see RA in the *Intro to Systems Thinking* folder on your disk)

Figure 7-10.
The Relative Attractiveness Infrastructure.

Description of Behavior

The infrastructure is initialized so as to generate a steady-state with respect to relative attractiveness. Each of the three components of attractiveness (density of development, unemployment, and housing availability) are set equal to the values for each component taken on by the town/city/region to which this population center is being compared—i.e., the "attractiveness gradient" is neutral. In time period three, a 10% influx of new businesses occurs, boosting jobs by the same percentage, and hence increasing attractiveness from an jobs standpoint. However, as Figure 7-11 shows, the increase in attractiveness from employment (2) is somewhat offset by a decrease in attractiveness from the higher development density (3), caused by the increase in the number of business structures.

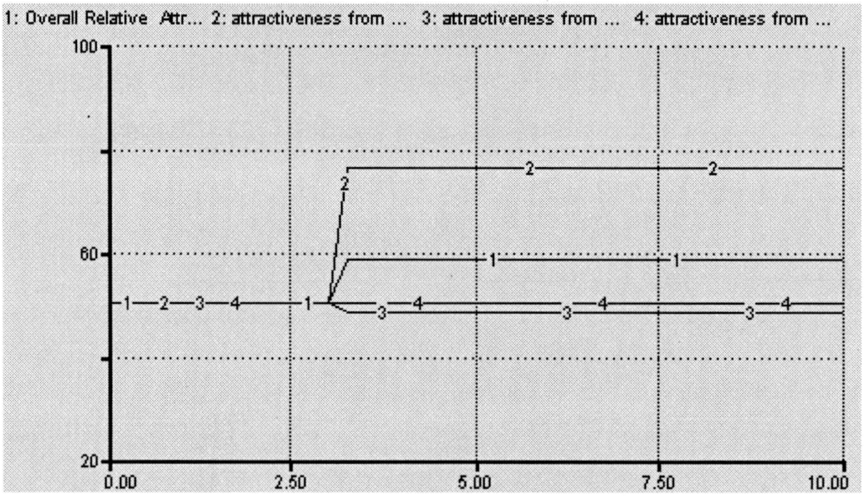

Figure 7-11.
A Shift in Overall Attractiveness Following an Increase in Jobs.

In interpreting Figure 7-11, it's important to note that the weightings on the three components of attractiveness are equal. That's why even though attractiveness from employment surged, the overall index of attractiveness moved far less (even without the offset due to increased development density). You might wish to play with these weightings, and then re-simulate the test to see how the impact changes.

Variations

Variations would consist principally of adding more attractiveness factors and allowing the weightings on those factors to vary (rather than remain fixed). One of the apparent characteristics of human perceptions processes is that, once a particular component of attractiveness is "satisfied," people tend to weight it less (i.e., to "take it for granted"). Weights could be set up to vary based on this notion—much like the way Maslow's hierarchy of needs works.

Part 2

The "Writing" Process

10,000 Meter, System as Cause, Dynamic, Scientific and Empathic Thinking

The chapters in this Part of the guide are designed to help you develop your ability to "write" compositions using the *STELLA* software. Chapter 8 begins with and overview of the "writing" process. Chapter 9 then provides an illustration of the process. Finally, Chapter 10 offers some "good practice" guidelines for executing the process.

The chapters in Part 2 will serve as useful reference materials as you construct models. If you get stuck somewhere in the modeling process, use the guidelines, illustrations, and examples presented in the chapters to get yourself back on track.

Chapter 8

An Overview of the "Writing" Process

It is fitting that we begin the "Writing"/Model-construction portion of this Guide by citing words of wisdom from four eminent scholars. You would do well to heed the sage advice dispensed in Figure 8-1.

"All models are wrong. Some models are useful."

Deming

"Seek simplicity...then, distrust it."

Whitehead

"The best explanation is as simple as possible...but no simpler."

Einstein

"Perfection is attained not when there is nothing left to add, but when there is nothing left to take away."

St. Exupèry

Figure 8-1.
Words of Wisdom for Guiding the Model-construction Process.

Each sage is essentially saying the same thing. In "American-ese:" Keep it Simple, Stupid...and, remember that when you do, it's not "true."

Why write? The most common reasons are to share one's thoughts and feelings, to entertain, to instruct/inform, and to incite to action. A not so commonly considered purpose is, to *learn*! But, in fact, for many authors, learning is a valuable by-product of writing experiences. And, in the case of "writing" a *STELLA* model, learning is not just a by-product, it's usually the *primary* reason for undertaking the effort. But it's not just the "writer" who learns as a result of a *STELLA* modeling experience. "Readers" of *STELLA* models also can be beneficiaries of a tremendous amount of learning—if the author writes the model in a way that facilitates such an outcome. So, there are some very solid reasons why the process of constructing a *STELLA* model should not be divorced from the process of learning from that model. Hence, I'll overview the steps in the "writing" (model-construction) process, and then discuss the progression of a parallel learning process.

The purpose of this Chapter is to provide a framework of "steps in the process" that will be used to frame the illustrative application you'll walk through in Chapter 9. Detailed guidelines for executing each step are then presented in Chapter 10.

Framing the Steps

First, I'll provide an overview of the "writing"/learning processes in diagram form. Then, I'll briefly describe each step.

Figure 8-2 diagrams the steps in the two parallel processes.

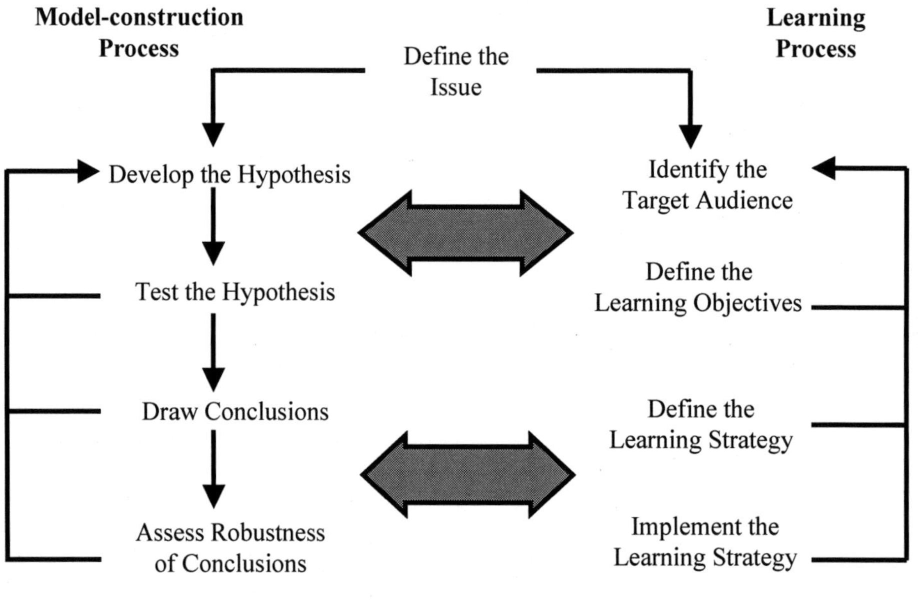

Figure 8-2.
The Steps in the Model-construction/Learning Processes.

As you'll discover, successful execution of the steps will draw upon the set of Systems Thinking skills that you began to develop in Part 1 (i.e., *Operational, Closed-loop, and Non-linear Thinking*). In addition, the five remaining Systems Thinking skills will be valuable in effectively executing the parallel processes: *Dynamic, 10,000 Meter, System as Cause, Scientific and Empathic Thinking.*

Spend a little time "processing" Figure 8-2 before proceeding with the text. You may even want to photocopy it to have available for reference when you pursue your model-building efforts.

As you can see by examining the Figure, the sequence of steps in both processes is far from linear. You will do a lot of iterating down, and then back up, through both sets of steps. The arrows that loop back represent learning. Note the large two-headed arrows that link the parallel streams. The intent of these arrows is to visually reinforce an important point: *although the streams run in parallel, there is a lot of interplay between them!* Each informs the other.

It is our strong recommendation that as soon as an issue is cast, you inaugurate a learning process to run in parallel with model development. The learning process depicted in Figure 8-2 is not your own, though certainly you will learn a lot through the process of constructing a *STELLA* model. Rather it is a process designed to ensure that others will learn from your "writing."

You may be a teacher/faculty member building a *STELLA* model to serve as an "engine" for a student exercise or a classroom lecture. You could be a researcher sharing findings with colleagues or funding sources. You may be an administrator who is looking to communicate a proposed solution across an entire organization. You might be a student who is tasked with "making available" what you've learned to other students, teachers, or parents. Whoever you are, if your *STELLA* model helps you to learn something, it's useful to make that learning available to others. That way, we all benefit from each other's development—and in the process, also feel good about sharing what we know.

Model-construction Process

I'll discuss the Model-construction ("Writing") Process steps first, and then treat the parallel Learning Process steps. I'm doing this for clarity of exposition purposes, not because that's the way the processes actually should be executed.

Define the Issue: Dynamic Thinking

Activity streams in *both* processes pirouette off purpose. Far too often, people embark upon model-building and/or learning activities with no clear, explicit statement of purpose to guide their efforts. Until you can state clearly and succinctly a purpose for your effort, please do not double-click the *STELLA* software! In addition to a written statement of purpose, it's also important to develop one or more graphs that chart patterns of behavior over time for variables that define the dynamic phenomenon you are interested in trying to explain (and perhaps modify). The ability to couch phenomena in terms of graphs over time relies upon a set of skills systems thinkers refer to as *Dynamic Thinking.*

Generally speaking, there are two broad categories of purpose for *STELLA*-based modeling efforts. The first is to create a "research tool." The second is to create a "learning tool." In principle, there is no reason why a single model can't serve both functions. In practice, however, this rarely occurs...and for good reason. Research tools tend to be "answer generators." The associated *STELLA* models usually are large, and place a premium on having highly-precise parameter values and generating numerically accurate results. By contrast, the distinguishing characteristic of *learning* tools is that they inspire insights—which is to say, they are capable of catalyzing changes in the assumptions comprising a mental model. In order to accomplish this end, they must remain small, and often contain only relative (i.e., internally-consistent) parameter values—as opposed to absolute and numerically precise ones. Yes, size *does* matter if what you are trying to do is "alter" the mental models people carry around in their heads. Small is *always* beautiful in this arena!

Over the 20 years or so that we've been constructing models, we've developed a distinct preference for developing learning tools first, and then, when needed, moving on to research tools. Not that the latter are less important. It's just that we've seen too many people head straight for the trees, never to emerge with any sense of a forest. Too often substantial time is invested, yet the resulting efforts end up literally and figuratively "barking up the wrong tree!"

In any case, whether your preference is for large research-oriented models, or small, learning-oriented ones, either effort should be guided by a sharply-focused issue, couched as a dynamic phenomenon to be understood.

Develop the Hypothesis: 10,000 Meter and System as Cause Thinking

Once a clear picture of the dynamic phenomenon being addressed is developed, the next step is to articulate a hypothesis that you feel can explain how the phenomenon is being generated.

There are three fundamental questions that must be answered in developing any hypothesis, or in "writing" any composition. The first is: *What should I include within the storyline/model boundary*? You are, of course, at the same time, answering the question of what to *exclude* (i.e., to allow to remain *outside* the model boundary). This first question is therefore a *breadth* question. The second question is: *In how much detail should I represent the elements I've decided to include*? This is a "character development," or *depth*, question. Figure 8-3 should help in visualizing these first two questions.

As the Figure is intended to suggest, a Systems Thinking spawned model tends to have a pretty broad extensive boundary, and a pretty shallow intensive boundary—though some elements may justify "going deep." Choices always must be made, and this is how, on average, they go when embracing a Systems Thinking perspective.

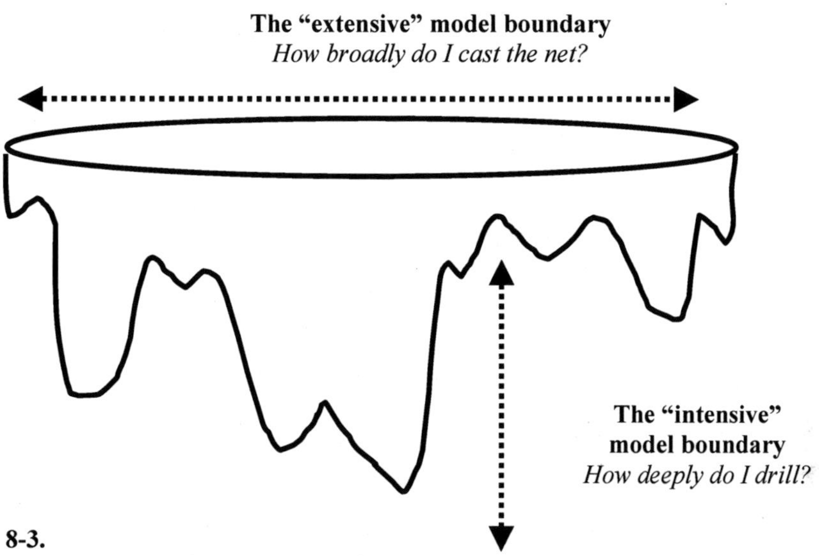

Figure 8-3.
Illustrating Extensive and Intensive Model Boundaries.

The third question that must be answered when developing any hypothesis is: *How will I represent the elements, and relationships between them, that I have chosen to include?* Chapters 2-6 addressed this question, by working to build your *Operational, Closed-loop, and Non-linear Thinking* skills.

The skills needed to help you answer questions one and two are: 10,000 Meter and System as Cause Thinking. These skills were defined in Chapter 1, where they were referred to as "filtering skills." They help filter reality so that the model can achieve the "as simple as possible, but no simpler" standard so eloquently articulated by Einstein.

Figure 8-4 helps explain why these two skills are needed. 10,000 Meter thinking helps to elevate your perspective, lifting you up out of the details and putting you on the "high road"—from which you can better see the "big picture" set of relationships responsible for generating a particular dynamic phenomenon. If you remain in "the depths," and try to expand a model's extensive boundary, it is very likely your modeling efforts will "not converge." By the way, it is also 10,000 Meter thinking that enables an insight gleaned within one discipline to become transferable, facilitating learning in other disciplines.

System as Cause thinking helps to enforce "Occam's Razor"—the simplest explanation that can account for the phenomenon is the *best* explanation. Remember back to the slinky in Chapter 1. System as Cause Thinking filters out any exogenous "drivers" of behavior as candidates for inclusion within the model boundary. Doing so helps to ensure that a sparser population of elements ultimately ends up within the confines of that boundary.

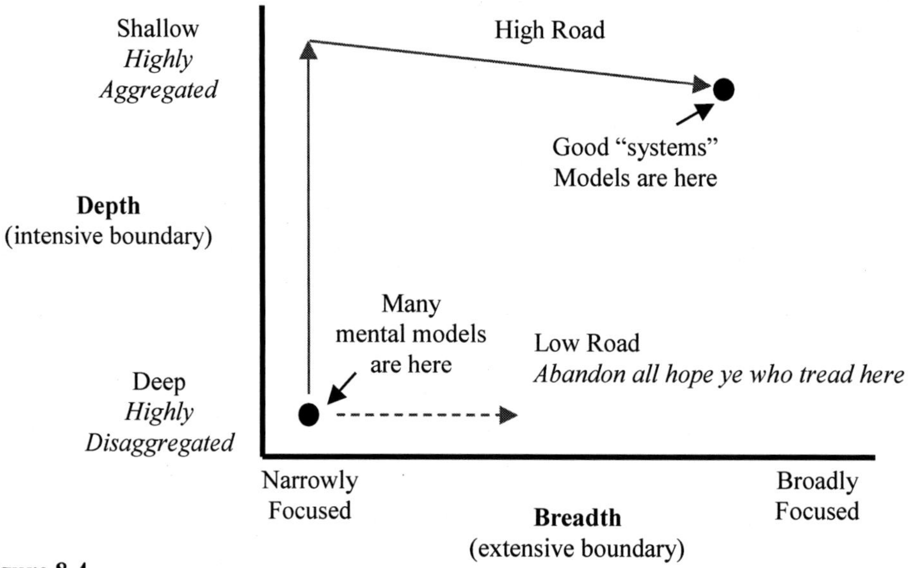

Figure 8-4.
The Typical Pathway to a Systems Thinking Model.

Test the Hypothesis

Once you've constructed a simulatable model, the next thing you'll want to do is test it. Certainly one test you will want to do is see if your model can replicate the dynamic phenomenon that inspired you to construct the model in the first place. But before jumping to this type of test, you'll first want to conduct a round of what are called "robustness" tests. Scientific Thinking skills guide this type of testing. You'll be placing your model in steady-state; changing one thing at a time; and then recording results—just as a scientist would do when conducting careful experiments in a lab.

Robustness tests are designed to make you aware of the limitations of your model. In essence, the question is: *Under what conditions does this model stop making sense*? Remember: All models are wrong.

That means, under some set of conditions, any model will stop making sense. You simply wish to verify that your model "holds up" well over the range of conditions within which you wish it to apply.

All robustness tests are conducted from steady-state using simple, one-time-change, test-inputs like STEP and PULSE functions (see the *Help Files* for more information on these built-in functions). The process of initializing a model in steady-state often reveals interesting things about the relationships between elements in your model. How to achieve a steady-state with your model is discussed in detail in an Appendix to Chapter 10. The notion of using "idealized" test-inputs to knock your model out of steady-state is that you just want to "ping" your model once, and then see how to reacts—as opposed to "driving it" with a time-varying input (in which case it becomes difficult to separate the response of the system from the input variation itself!).

Robustness tests subject your model to "extreme-condition" shocks. Examples of "shocks" would include: pulsing out 90% of what's in a stock; stepping up the volume of an inflow by 50%; these kinds of things. In each case, you are looking for "interesting" responses. Does the system return to steady-state following its response to the shock? Does something collapse, or go into an expanding oscillation? Does a key variable seek a new steady-state level? These sorts of tests help to build confidence in your model's formulations, and also make you aware of its limitations. In addition, they sometimes reveal high-leverage points—places where a little tickle gets a big reaction!

Again, throughout the regimen of Robustness Tests, you are exercising Scientific Thinking skills. Administer a simple one-time-change test; carefully record the system's response; proceed to next test. Such skills are vital to squeezing the maximum amount of learning out of your simulations. More on learning in a moment.

Once the model has successfully undergone a regimen of robustness tests, the next round of testing focuses on replicating a Reference Behavior Pattern (RBP). An RBP is the graph(s) you produced when you defined your issue. If the model successfully replicates your RBP, it means you have an "entertainable hypothesis" on your hands. It does not mean your model is "valid"—*no matter how precisely model-generated behavior tracks the reference behavior*! It just means you have an explanation that *can* account for the phenomenon, not that it *does* account for the phenomenon.

Draw Conclusions

Once you have an entertainable hypothesis to work with, the next step in the process is to use it to draw some conclusions. These may be about why the phenomenon is unfolding as it is, what actions might be taken to alter that unfolding, what unintended consequences might be set in motion by intervening in the phenomenon, or about other things to do with the phenomenon.

Assess Robustness

Having drawn some conclusions, the next step in the process is to understand just how "robust" they are. Under what scenarios (variations in *external* conditions) do the conclusions hold, and under which do they crumble? In addition, you'll want to determine how sensitive your conclusions are to variations in the values of *internal* parameters. For example, understanding the range of external and internal conditions under which any initiatives you might take remain "the best course of action," enables you to be proactive about adapting to "outside the box" circumstances should these arise.

Learning Process

Let's now turn full attention to the Learning Process. It's worth saying again. This process is not to be conducted in series with the model-construction process. It runs in parallel, and should begin as soon as the issue is cast.

The learning process is defined by the answers to three questions: *Who's the audience*? *What are they supposed to learn*? *How are they going to learn it*? The first two are straight forward to answer. The third requires a bit of discussion.

Identify the Target Audience

Possible audiences include: students, colleagues, clients or sponsors in a research/professional context, and the public (in one of its many guises). Different audiences usually will require different learning strategies, though it is our experience that most audiences prefer *active* to *passive* learning strategies.

Define the Learning Objectives

The primary focus of the "what are they supposed to learn?" answer usually is some substance-specific material. For example, if students are the audience, you might want them to understand what Chaucer's *Canterbury Tales* is really all about, how Newton's Second Laws work, or why populations so often overshoot their carrying capacity. In some cases, there is a *secondary* learning focus; one that is substance-independent: the so-called *critical thinking skills.* This is where Systems Thinking lives. In this arena, the objective is development of skills, understandings, and insights that transcend a particular discipline or research context. As you are about to see, consciously using Systems Thinking as a vehicle for boosting learning productivity within a substantive arena is a strategy that can pay very substantial dividends.

Define the Learning Strategy

The final question—*how are they going to learn it*—is the most interesting of the three, both because its answers come in the most flavors, and because the way in which the question is answered will determine how productive the learning experience will be.

A major cleaving of the universe of learning strategies occurs when you make the choice between a *passive* and an *active* learning strategy. The former features lectures and textbooks. People "sit and listen," or "read and underline." They are spoon-fed information. By contrast, when an *active* learning strategy is employed, *discovery* is always the essential ingredient. Learners have to *participate* in order to learn.

Teachers who embrace an *active* learning strategy do their work "in advance" by creating a fertile soil learning within which learning can bloom. While learning is blossoming, the teacher, if present, acts as a catalyst—keeping learning on track, and reinforcing fruitful learning initiatives. Oftentimes, particularly when the *STELLA* software is employed, the teacher will have disembodied their expertise as a learning catalyst—something that becomes essential in a distance-learning environment. The teacher's presence now manifests within the *STELLA* model in the form of "built in" coaching sequences that re-direct learners when they get off track, or stimulate learners' thinking with a provocative question.

Clearly, the *STELLA* software supports *active* learning strategies. In a teaching context, *STELLA*-based exercises range along the continuum illustrated in Figure 8-5.

Figure 8-5.
The Continuum of STELLA-based Exercises.

At the far right of the continuum are exercises in which students are asked to *construct* models from scratch. At the far left, students "tweak" a model someone else has constructed. Included in this category are "Flight Simulators" (with built-in coaching), and "Virtual Laboratories." In the middle of the continuum are exercises in which students *extend* an existing model by adding elements and relationships.

Whether an exercise employs constructing, extending, or merely exercising an existing *STELLA* model, the purpose is always to surface the underlying mental model. Students are led to examine the assumptions that constitute these models—both the specific substantive content of the assumptions, as well as the *general nature* of the assumptions. In so doing, they build both their content-specific understanding, and their general capacity for developing understanding—two of the stocks, as suggested in Chapter 1, that must be filled if we are to produce students capable of addressing the challenges of interdependency that will characterize life in the new millennium.

An important synergy often is tapped when employing a *STELLA*-based learning strategy. Models that allow students to actively and creatively experiment with different personalities for Hamlet, or different acid rain scenarios for an ecosystem, or different interest rate policies for stimulating an economy, often ignite a student's imagination. Not surprisingly, when this occurs, motivation receives a dramatic boost and learning productivity increases. Students learn faster *and* develop a deeper, richer level of understanding. This buoys confidence and stimulates interest in continuing to learn. And because whether it's Hamlet, an ecosystem, or an economy, they'll see stocks and flows, feedback loops, delays and non-linear relationships, the rigid dividing lines between disciplines begin to dissolve. The ability to think *horizontally* is honed. Systems citizens are cultivated.

One particularly exciting learning strategy that is slowly but surely gaining increasing favor is "students teaching students." In this approach, students are charged with making what they've learned available to other students. The typical form a "students teaching students" strategy takes, often goes as follows...Students construct a *STELLA* model to shed light on some issue, theme, or set of concepts, in a given substantive discipline. They then are charged with designing an interface for their model, and usually also some built-in coaching sequences, that will allow other students (unfamiliar with the model, and likely also with the *STELLA* software), to learn something substantive from exercising it.

The volume of learning for the students creating these *STELLA*-based "learning environments" is *enormous*! The first course of learning occurs during the process of constructing and exercising the *STELLA* model. But because the students doing the constructing are obligated to share what they have learned with other students (who did not participate in constructing the model), a *second* course of learning occurs. This course is even more delicious than the first because it exercises perhaps the most "system-sy" of the Systems Thinking capacities: *Empathy*, the ability to experience as one's own, the feelings of another.

To be able to create "fertile learning soil" for someone else, a person must be able to "put themselves in the others' shoes," and to "see the world from their point of view." These are the real, organic "systems" capacities. They are capacities that can't be learned by reading a book. They also are the capacities that enable real communication to occur, and that breed tolerance for differences. They are the capacities we vitally need in order to remain viable as a species in the new millennium.

Things get very interesting in the "students teaching students" domain when the students for whom the fertile learning soil is being created do not live next door, but rather on the other side of the globe. Now the empathic bar is raised to a truly challenging level. And this becomes easy to achieve using the Internet. Just imagine the possibilities!

As intriguing as student-focused learning strategies may be, students are not the only people who learn. Faculty and researchers, for example, must share their results with colleagues, the public, and funding source administrators. The *STELLA* software makes available a nice learner-directed sharing strategy to address this learning need. It's called "storytelling," after the software feature of the same name.

STELLA's storytelling functionality allows the logic of a model to be unfurled one "chunk" at a time—with what constitutes a "chunk" being defined by the person who creates the storytelling sequence. Each chunk also can be annotated with text, a graphic, a sound, or a video—allowing, once again, for the effective disembodiment of pedagogic expertise. The rate of unfurling is under the complete control of the learner. Each chunk, or any combination of chunks, can be simulated to see what dynamics it produces. This enables someone "processing" a *STELLA* model to build up their understanding chunk-by-chunk, and to do so at their own pace, and in a discovery-oriented way. All in all, an extremely powerful way to enable colleagues, clients, and the public to "bring themselves up to speed" without being "lectured to."

Implement the Learning Strategy

Once a learning strategy has been defined, it must be implemented. And then learning outcomes must be monitored so you can increase the effectiveness of the associated learning strategy over time. The monitoring and improving processes for *STELLA*-based curriculum materials are not really any different than those you'd employ for any such materials. But there is one species of monitoring and improving that assumes a unique form when the *STELLA* software is involved. That's the monitoring and improving associated with an *organizational learning strategy*.

A growing number of organizations, including many educational institutions, have taken up the challenge of becoming "learning organizations." As noted in Chapter 1, this term has not been operationally defined to my satisfaction as yet. But clearly, a sufficient number of organizations have become intrigued by whatever they perceive it to mean that there is indeed an "organizational learning" movement going on out there. So, let's spend a minute looking at what the *STELLA* software can bring to this party.

The software's principal contribution in the realm of organizational learning is that it serves as a tool for enabling construction of highly-accessible receptacles for the collective understanding and insight developed by the members of an organization. A good name for the collection of receptacles is a *STELLA*-based "organizational learning infrastructure." Such an infrastructure would house more than just

STELLA models, but here we'll focus only on this component. Central to the *STELLA*-based infrastructure is a server-based inventory of *STELLA* maps and models that would be maintained and operated much like a reference library. At any time, anyone with access to the repository (which presumably would be any member of the organization) could download a particular model, review the associated assumptions, run simulations to test initiatives or conduct what-ifs, and also to *propose modifications*. A formal process would be established for reviewing all proposed modifications before any updates to the "golden master" were implemented.

Over time, through this "ongoing review process," the degree of alignment in underlying mental models across the organization would increase. More people would "be on the same page," facilitating execution of any initiative being implemented by the organization. In addition, the *quality* of the models in the repository would be systematically ratcheted upward over time as feedback from reality weighed in, and was then used to re-tool model assumptions. And, as the quality of the *STELLA* models improved, so too would the quality of the associated mental models. *Real* organizational learning would come to pass.

Embarking upon a model-construction process, without giving adequate thought to an associated a learning process, runs the risk of squandering a *huge* potential for learning! Do yourself, and the rest of the world, a favor. Before you begin constructing any model using the *STELLA* software, take some time to consider who an appropriate audience might be with whom to share the fruits of your labors. Then, be explicit about what it is you want to share. And finally, identify what you feel will be an effective *active* learning strategy.

What's Next

You now have an overview of the processes for using Systems Thinking and the *STELLA* software to both construct a model and to share the resulting learning with others. In the next Chapter, I'll walk you through these parallel processes using an example taken from the Life/Environmental Sciences. The example is easy to understand, so that no Life Science expertise is needed, or assumed. Then, in Chapter 10, I'll offer a cookbook of guidelines and principles for executing each step in both processes.

Chapter 9

Illustrating the "Writing" Process

There are many "products" of the "writing" process that accompany the software package. See especially the *Getting Started Guide* available on the CD that houses the software. I suggest that you examine several of these products to see how they're constructed, and to learn something about the things that will not be illustrated in this Chapter—such as "built-in" coaching (coaching sequence guidelines are provided in Chapter 10).

I have chosen development of an "extension" exercise (recall the continuum presented in Chapter 8, Figure 8-5) to illustrate the "writing" process. Extension exercises are hybrids, located between exercising and constructing, and thus are useful for illustrating aspects of both processes. In addition, even for those new to the *STELLA* software, developing an extension exercise is a reasonable undertaking because the departure point is a well-constructed model nucleus. Extension exercises are also good for students. Rather than having to begin with a blank screen, and being faced with making difficult decisions about model boundaries, they'll instead be asked to make marginal additions to an already well-constructed, well-bounded core model. This enables students to view "good practice," and then to replicate that practice by "adding on to." The burden on teachers and faculty members also is lightened, as they aren't obligated to try to make meaning out of a class-load of free-form model-construction efforts.

I'll use the "steps" in the process outlined in the previous chapter to frame the discussion. Those steps are reproduced in Figure 9-1 for your convenience.

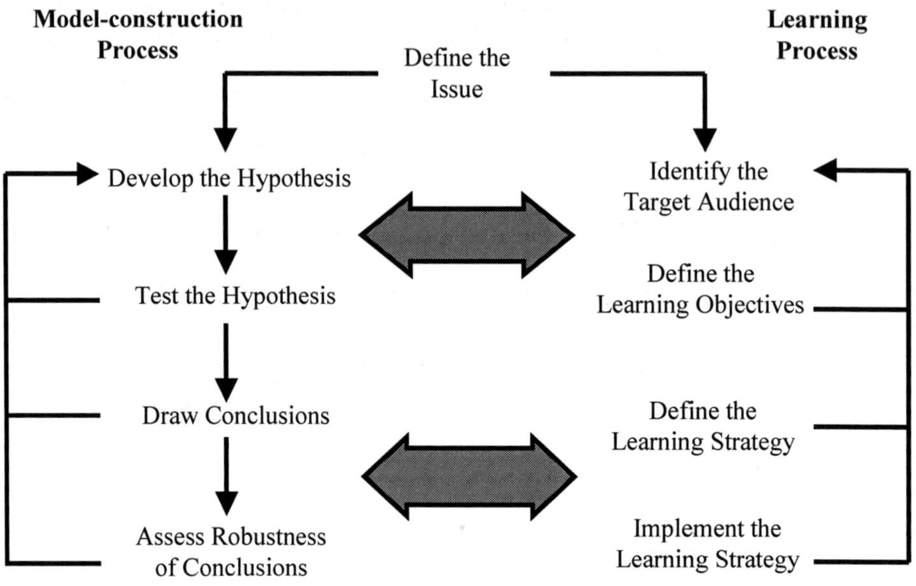

Figure 9-1.
The "Steps" in the Model-construction/Learning Processes.

In this Chapter I'll begin by defining the issue. Then I'll walk through the steps in the learning process. Finally, I'll work through the model-construction steps.

Defining the Issue

The issue for this illustration is taken from Biology. The topic is the evolution of a trait in a population under the pressure of natural selection. "Natural selection" simply means that certain traits tend to give individuals who possess them better chances of surviving, and hence of passing on those traits to their offspring. As a result, over time, more and more members of the population will tend to possess these traits. You will see in more detail how this process works in the illustration. If you had no prior exposure to the concept of natural selection, you now know enough about the basics to appreciate the illustration.

The substantive purpose of the exercise is to enable students to develop an experiential understanding of natural selection as an evolutionary force. There also are secondary, critical thinking skill, purposes to be achieved. The exercise is intended to reinforce the idea that "all models are wrong," and that the way to discover where they're wrong is to challenge their boundaries. Another general thinking skill to be developed is the role reinforcing and counteracting feedback loops play in determining the dynamics a system exhibits. Finally, the exercise is intended to provide practice in employing the "scientific method," which systematically advances understanding by changing only one thing at a time.

Figure 9-2 depicts a Reference Behavior Pattern for the issue. As you can see, the model that students must extend should be capable of generating an evolution in the average speed of a rabbit population over a forty-year period. The amount of the increase is about 50%.

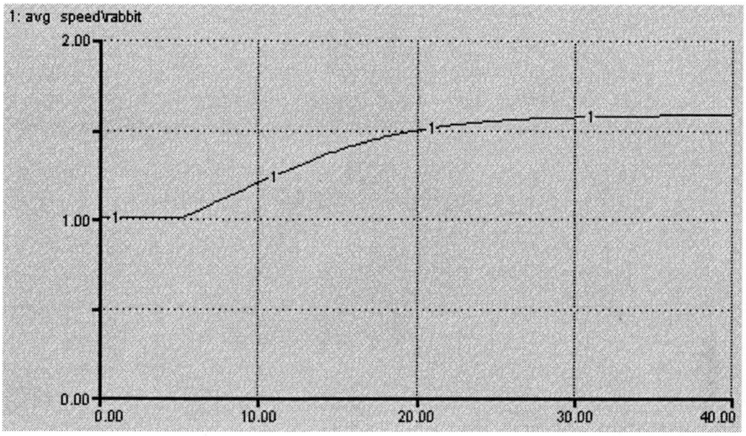

Figure 9-2.
The Evolution of Speed in a Rabbit Population.

Identify the Target Audience

The target audience for this modeling effort is undergraduate students in an Introductory Biology course. The audience will have had some exposure to the *STELLA* software, but by no means be expert in its use—or in the Systems Thinking skills that support its use.

Define the Learning Objectives

As already noted, in this instance, learning objectives are both substantive (understanding how natural selection works) and substance-transcendent (develop critical thinking skills). In the latter case, the intention is that students gain more familiarity with the language of stocks and flows, further develop their grasp of feedback loops, and build their appreciation for the power of the scientific method as an approach to gaining understanding.

Define the Learning Strategy

The learning strategy is to pursue an *active* learning approach using the *STELLA* software to create an "engine for learning." From the continuum of *STELLA*-based exercises, (which includes exercising, extending and constructing), the choice is "extension." The rationale is that it's important, early-on in a student's formal model-building career, to ensure that they learn from "good practice."

Conceptualizing, and then constructing a model from scratch is an extremely challenging endeavor—principally because it entails making choices about model boundaries (both extensive and intensive). Extension exercises provide a "good start," a nucleus that's well-conceptualized, formulated, and numerated. Students have something scoped and solid to build on. They are permitted to work on the exercise in teams of up to size three. Larger teams encourage "free riders;" smaller teams need more diversity of viewpoint.

Implement the Learning Strategy

Implementing the learning strategy consists of first working through the model-construction/testing process that students will be asked to go through. The resulting exercise is then constructed by simply re-tracing the progression of intermediate model forms that lead up to the "final" version of the model.

The "starting point" model looks like what appears in Figure 9-3.

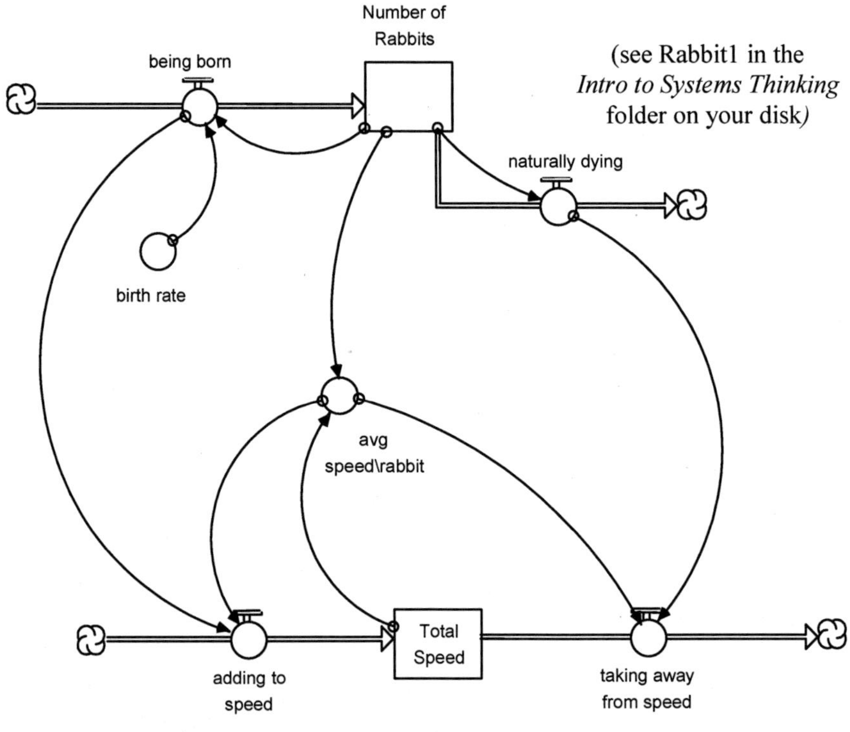

Figure 9-3.
The "Starting Point" Model.

The notion in the starting point model is to provide students with the "analytically tricky" piece of modeling—the calculation of the moving average of the attribute that they will attempt to make evolve (i.e., average speed). Doing so, keeps students from spending time spinning their wheels on something that really is a "useful modeling trick" rather than a substantive, or substance-transcendent, learning. Students have in the core model an intuitive structure that's well-formulated, and numerated in an initial steady-state.

Students are provided with the following directions for completing the assignment...

> Your challenge is to add a fox population to the core model structure depicted below [this would be the diagram that appears in Figure 9-3], such that a natural selective pressure is created that causes the average speed of the rabbit population to *increase* over time. The increase should be in the vicinity of 50% over a 40-year period as illustrated in the graph [this would be Figure 9-2]. Do not spend time trying to track the graph exactly! Just seek to generate a pattern that shows an increase in magnitude of roughly 50% over the specified time horizon.
>
> To simplify your work, do not allow the fox population's speed to evolve, or for that population to vary in magnitude.
>
> It's important to save any intermediate models that you create in the journey from "starting point" model to final model. For each (there should be at least one, and preferably two or three!), explain the shortcoming that inspired you to continue modeling, and how the next version of the model addresses the shortcoming.
>
> For the final version of your model, be certain to explain what is causing the dynamics it is generating. For example, what causes the average speed to cease increasing, and how would you cause it to rise to a higher average value in steady-state?
>
> The first thing you should do is to make a Base Case simulation in which you verify that the model is initialized in steady-state. Be sure you understand how steady-state is being maintained in the starting point model.

The two simplifying assumptions stated in the assignment are crucial to making the exercise achievable for beginning-level *STELLA* users. If the fox population's speed evolves along with the rabbit population's speed, students would have to come up with some other way to limit "the footrace" that will ensue (i.e., rabbits get faster, causing foxes to get faster, causing rabbits get even faster, and so forth). And, if the fox births and/or deaths were tied to the availability of rabbits, we would pretty quickly get into some fairly complex two-species population dynamics. In developing extension exercises, it's essential to go through the steps in the extension pathway students are likely to follow, and to ensure that the required additions to the model lie within the capability of the audience for whom the exercise is being designed.

We'll now proceed with the steps in the Model-construction Process in the manner of a student working on the exercise.

Develop & Test the Hypothesis

The student's first task is to develop and render a *STELLA*-based hypothesis that can account for the *Reference Behavior Pattern*; i.e., the 50% (or so) increase in the average speed of the rabbit population over a forty year horizon. We've been told that foxes will provide the natural selective pressure on rabbit speed in this ecosystem. In this case, the *one* attribute undergoing a natural selective pressure is the foot speed of the rabbit population. The idea is that foxes will catch, and eat, the slower-afoot rabbits. This will leave the more fleet-footed rabbits around to reproduce. The offspring will have the fleet-footed attribute, and then they'll reproduce. And so, as a result of introducing a predator (i.e., foxes) into this ecosystem, the average speed of the rabbit population will increase over time.
The first step in *extending* the base model is to add a fox population. Recalling the two simplifying assumptions, Figure 9-4 shows the model with the foxes included.

(see Rabbit2 in the *Intro to Systems Thinking* folder on your disk)

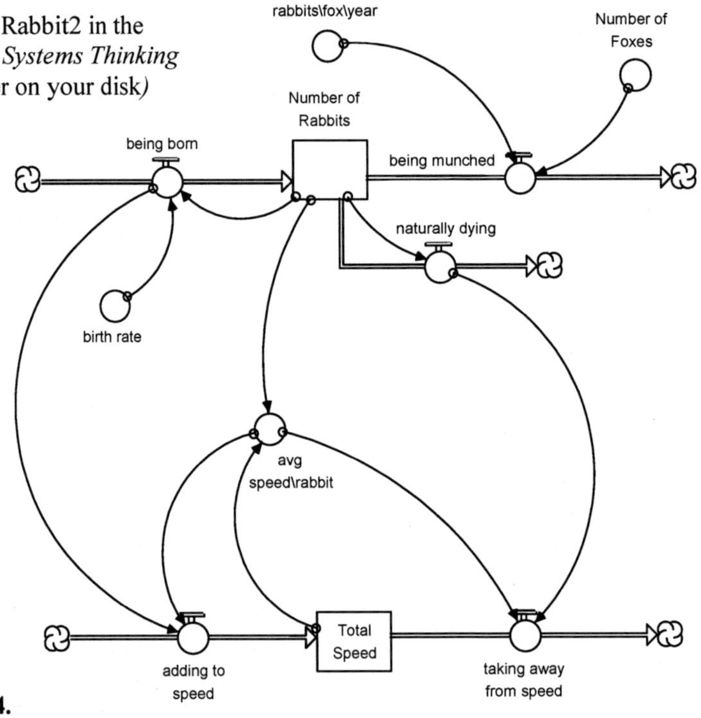

Figure 9-4.
The Model with Foxes Included.

The fox population is represented using a converter, even though in concept it is a stock, because we are not concerned with fox births and deaths (in fact, the number of foxes will remain constant). A *being munched* outflow from the rabbit population has been added, and formulated using the *external resource* template (see the Appendix to Chapter 5).

Once the new structure is added, the next task is to numerate the addition so we can see what impact it has on the system's dynamics. The guiding numeration principle at this point in the process is the idea of steady-state. We will want to choose numbers that leave the system sitting in a balanced state. Because a second outflow has been added, we'll need to allow it to carry some of the total outflow from the rabbit population—now being completely carried by the flow *naturally dying*. We have lots of room to make numerical choices here. The reasoning I used went as follows…

I'll begin by picking a value for *Number of Foxes*. Rabbits are set to 500. I suspect that there would be far fewer foxes in the ecosystem than rabbits. Why? Because it takes many rabbits to feed a single fox for a year. I'll choose a value of 25 for the fox population. Could I have chosen 50? Sure. 100? Maybe. 250? Probably not. Not being an ecologist, I don't know what a reasonable steady-state balance between this predator and this prey would be. But it raises a very good question! And that's one of the side benefits of asking students to numerate models to achieve a steady-state. It raises very good questions! For now, I'll go with 25 foxes and see how well it works.

With 25 foxes, and a *being munched* flow that is defined as *Number of Foxes* times *rabbits\fox\year*, I need to choose a value for the latter parameter that yields a value for *being munched* that is something less than 125 (because 125 is the value of the *being born* inflow, and I need to save some of the total outflow for *naturally dying*). I will choose a value of 4 for *rabbits\fox\year*. That will yield a *being munched* volume of 100 (25 foxes times 4 rabbits\fox\year). Hence, the *naturally dying* outflow must now take on a value of 25, in order that the sum of the two outflows equals 125—which, as you'll recall, is the value of the inflow. This means we must go into the *naturally dying* outflow and change the death rate (a parameter that is "buried" within the *naturally dying* equation, rather than being extant on the diagram) from its current value of 0.25, to a value of 0.05. Making this change will cause the *naturally dying* flow to take on a value of 25 (i.e., 500 rabbits times 0.05 rabbits/rabbit/year).

Okay, now we're ready to see what the changes we've made do to the dynamics the system will exhibit. Figure 9-5 depicts a run of the modified model. As you can see from scrutinizing the graph, the good news is that the rabbit population remains in steady-state. The bad news is that the average speed per rabbit is now heading for the stratosphere! Why is that happening?

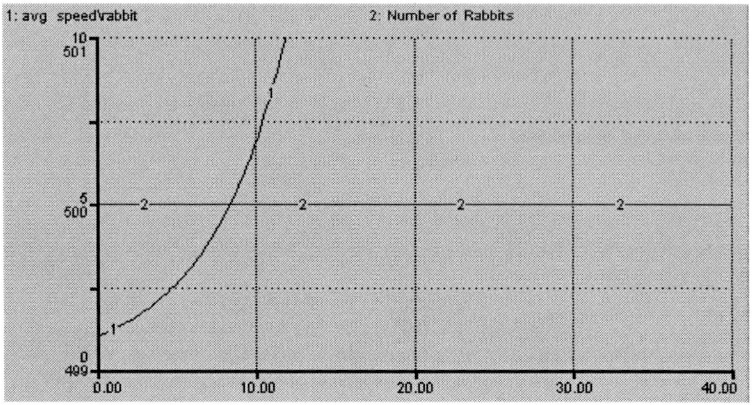

Figure 9-5.
Simulating Model Number Two.

One of the best ways to find out is to put a bunch of relevant variables into a table, simulate, and then look at the resulting numbers. Relevant variables in this case would be those having to do with the average speed calculation: *Total Speed, adding to speed, taking away from speed.* You'll find the numerical values for these variables displayed in Figure 9-6. As a look at the values indicates, rather than the *Total Speed* remaining constant, which it would have to do in order to have average speed remain constant, the stock is *increasing* in value! It's doing that because the inflow to the stock is greater than the outflow. And since we did nothing that would impact the inflow, the issue here must be with the outflow (*taking away from speed*).

Time	Total Speed	adding to speed	taking away from speed
.00	500	125	25
1.00	608	152	30
2.00	739	185	37
3.00	898	224	45
4.00	1,091	273	55
5.00	1,327	332	66
6.00	1,613	403	81
7.00	1,960	490	98
8.00	2,382	596	119
9.00	2,896	724	145
10.00	3,520	880	176
11.00	4,279	1,070	214

Figure 9-6.
Average Speed Variables.

A little thought, and further examination of the diagram reveals the problem. We've shifted most of the outflow volume from the stock of rabbit population to the *being munched* flow…but we have not wired the *being munched* flow up to the *taking away from speed* flow! As a result, the majority of rabbits who are dying, are, in effect, taking with them a speed of *zero*! The result is that the average speed of the remaining rabbit population is soaring to infinity! It's easy to fix this problem. A modified version of the model, with appropriate wires included, is shown in Figure 9-7. A simulation of the model is shown in Figure 9-8.

(see Rabbit3 in the *Intro to Systems Thinking* folder on your disk)

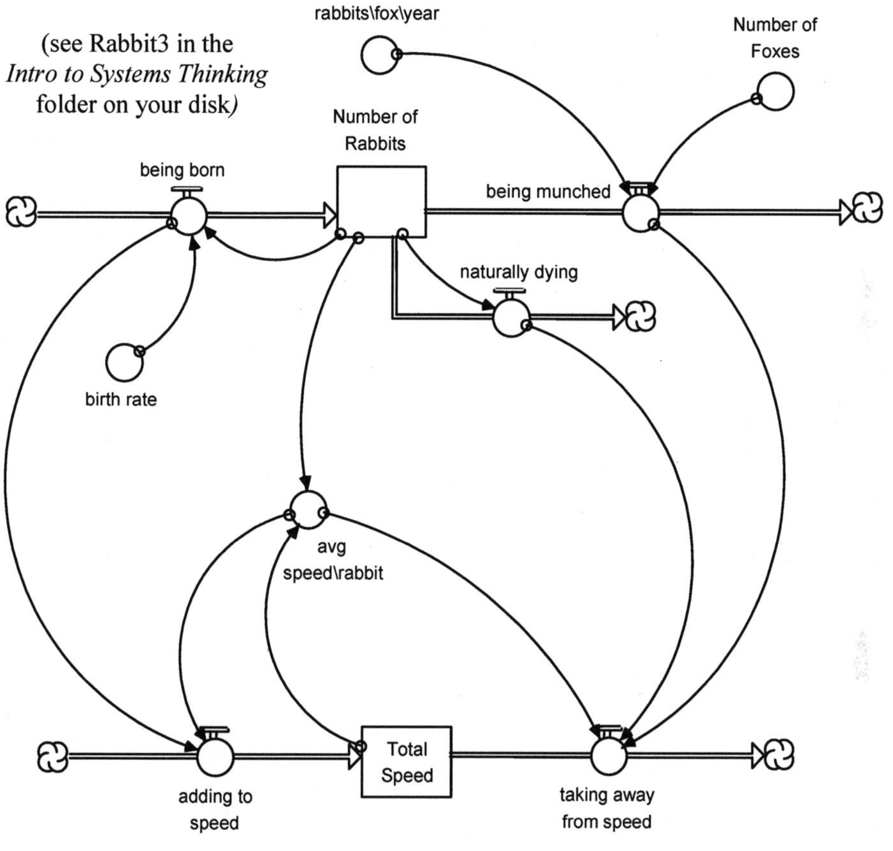

Figure 9-7.
Model Number Three.

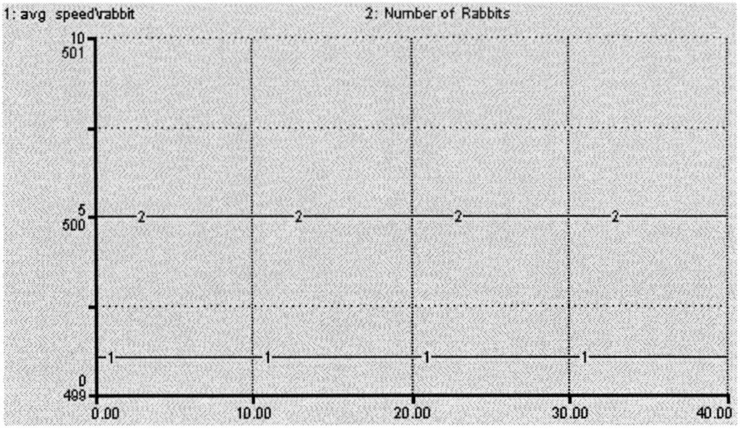

Figure 9-8.
Simulating Model Number Three.

As you can see from looking at Figure 9-8, the fix indeed fixed the issue of average rabbit foot-speed soaring off the graph. The system is now in steady-state.

The question now is how can we cause the average speed per rabbit to "evolve?" The "crazy" simulation results we produced before the *being munched* flow was linked to the outflow from *Total Speed* actually might be of some help in figuring out how to bring this about.

In that simulation, average speed was "evolving"—albeit for the wrong reasons. But, what we learned was that in order for average speed to increase, the rabbits that are being *munched* have to carry with them something *less than* the average speed of the population. They can't carry a zero—as they were in effect doing before the link from *being munched* was added. However, they can carry some value *less than* the average speed of the overall population. And, indeed that is just what natural selection theory says will happen: i.e., slower rabbits will get *munched.* We can include this logic by adding a variable to the model that represents the average speed of the rabbits that are being munched—separating it from the average speed of those rabbits dying naturally (who are assumed to carry with them the average speed of the overall population). Figure 9-9 shows the addition.

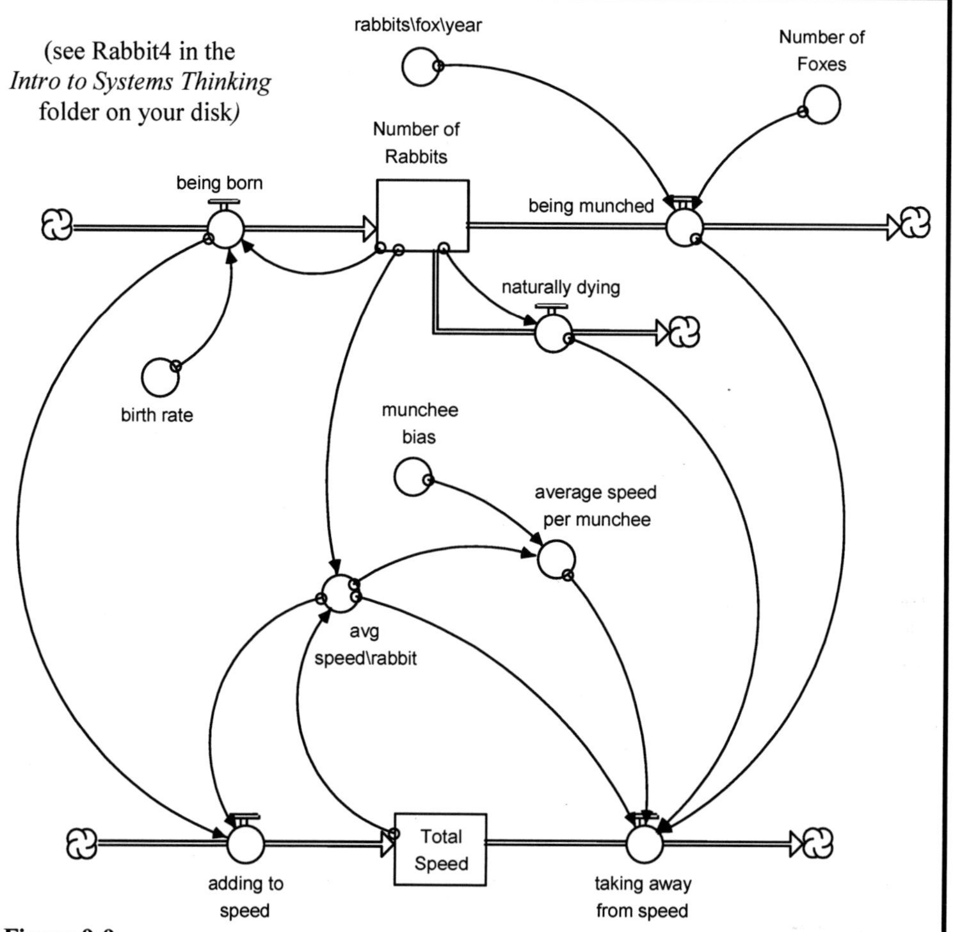

Figure 9-9.
Model Number Four.

The new variable *average speed per munchee* represents the speed the slower rabbits who are munched by foxes carry with them. *average speed per munchee* is calculated by multiplying the *avg speed\rabbit* by the *munchee bias*—the latter variable ranges between 0 and 100, values that reflect the percentage amount slower afoot the munchees are than the average speed of the population as a whole. The *munchee bias* is set to a value of 20, meaning that the rabbits who are caught by foxes are 20% slower than the average rabbit in the population. Figure 9-10 displays the new simulation results.

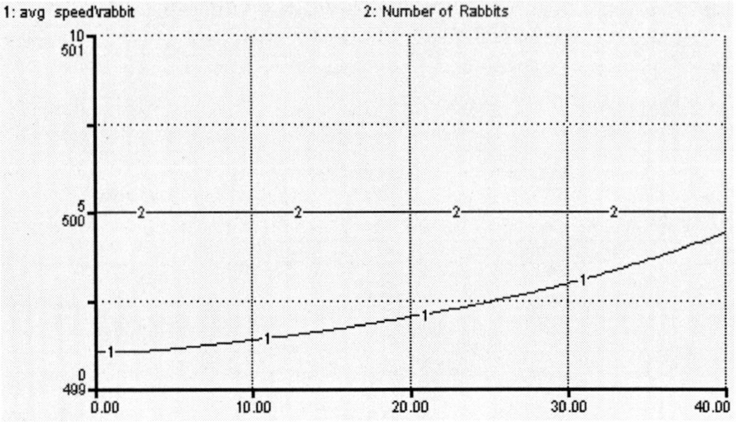

Figure 9-10.
Simulating Model Number Four.

The results are encouraging. The rabbit population holds its steady-state value. The average speed of rabbits in the population increases. The problem now is that the average is continuing to climb, and by the end of the 40-year horizon is almost 500% (rather than 50%) higher than its initial value. In addition, the average speed is not leveling off. But, perhaps the average speed will level off if we extend the time horizon, and maybe we just have too aggressive a "munchee bias." Easy to check…

If you set the length of the run to, say, 80 years (in the model "Rabbit4"), you will see that, alas, we've got a real problem—*not* just an overly aggressive "munchee bias." The average speed continues to grow, and to do so exponentially, such that after 80 years it stands at a value more than *twenty-one times* the initial value! So, it's back to the drawing board. Why does the average continue to climb?

Once again, a good approach is to throw the relevant variables into a Table, or onto a Graph page, and simulate to see what they are doing. Relevant, again, are variables having to do with rabbit speed. The new variable in this arena is *average speed per munchee*…we'll want to be sure to look at how it is changing over time. When we do (if you're interested, put the speed variables into a Table in "Rabbit4" and run a simulation—and, if you do, change the report interval within the Table dialog to 1.0), we discover that it begins at 1.0 (the average speed of the population) and then, once "turned on" at time 3.0, it immediately dips to 0.8 (20% lower than initial), and then begins climbing. By the end of the simulation, its value is more than 3.5! This means that even though fox foot-speed is *not* increasing, they are somehow able to catch rabbits—at the same, constant rate of four per fox, per year—who are more than *three and a half times* faster than the rabbits they were catching at the outset of the simulation. This doesn't make any sense!

Aha! The problem is that as average rabbit foot-speed increases, given a constant foot-speed for foxes, the number of rabbits munched per fox must *decrease* rather than remain constant. This idea is easy to implement into the model using a graphical function. Figure 9-11 shows the next incarnation of the model.

(see Rabbit5 in the *Intro to Systems Thinking* folder on your disk)

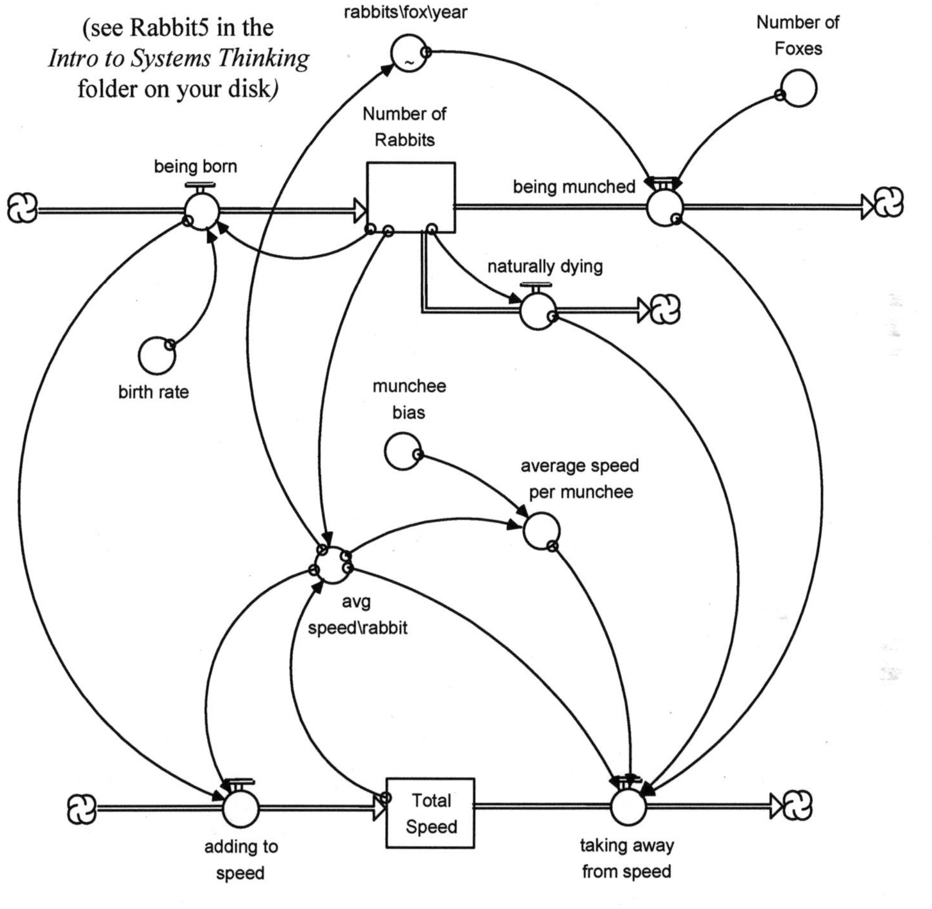

Figure 9-11.
Model Number Five.

As the Figure shows, *rabbits\fox\year* is now a graphical function that depends on the average speed of rabbits in the population. The graphical function relationship is displayed in Figure 9-12. As the Figure indicates, when the average speed of the rabbit population is 1.0 (the initial steady-state value), the average number of rabbits killed per fox per year is 4.0 (our initial choice of a constant value). Then, as average rabbit foot-speed increases, fox bounty declines, and vice versa.

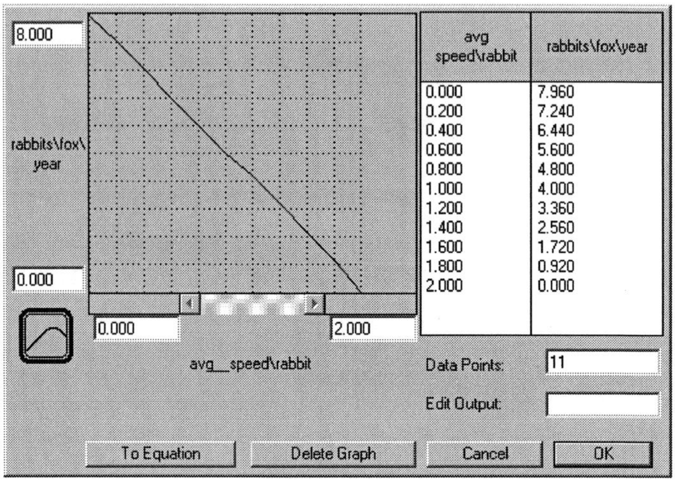

Figure 9-12.
Rabbits Munched Per Fox as a Function of Rabbit Foot-speed.

Figure 9-13 shows simulation results produced by the new model.

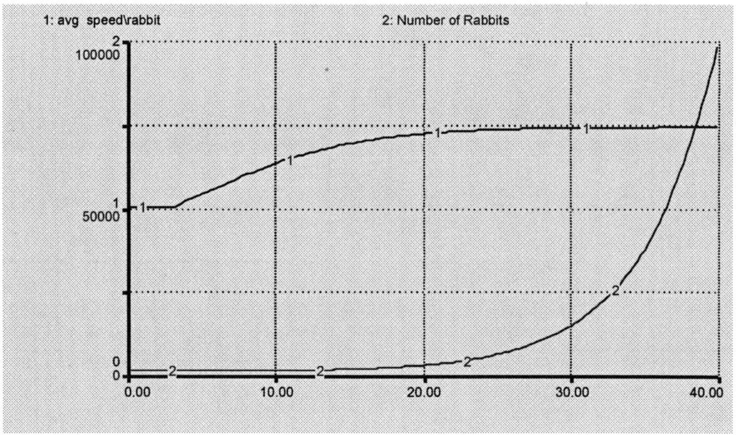

Figure 9-13.
Simulation Results from Model Five.

Well, now we have some really good news…and some modestly bad news. The former is that average speed per rabbit now increases by about 50% over the forty-year horizon and does so *asymptotically*—i.e., it reaches a limit. The bad news is that now the rabbit population is exploding!

At this point, the requirement stated in assignment has been met. A fox population has been added. As a result, the average speed of the rabbit population has climbed to a new steady-state level, roughly 50% above its initial value, and then leveled off. The fact that the rabbit population now appears to be growing without limit is "interesting," but does not necessarily need to be addressed by the student. What is causing the rabbit population to explode is certainly good grist for a classroom discussion.

To get a handle on why the rabbit population is exploding, it's useful to graph the inflow to (*being born*), and the two outflows from (*being munched* and *naturally dying*), the rabbit population. These flows appear in the graph depicted as Figure 9-14.

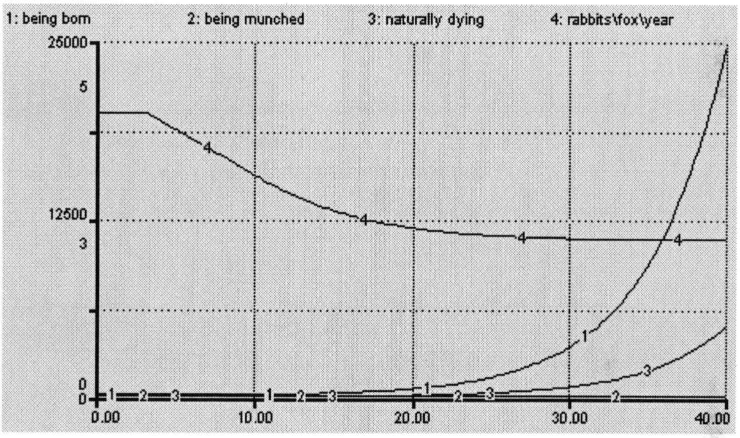

Figure 9-14.
Rabbit Flows and Rabbits\Fox\Year.

The Figure shows that the *being born* flow (the inflow to the rabbit population) comes to completely dwarf both the *being munched* and the *naturally dying* flows (the outflows from the stock). In fact, the inflow is growing exponentially, while the *being munched* flow appears to be remaining relatively constant (and is actually declining slightly). The *naturally dying* outflow also is growing exponentially, but at a much slower rate than the *being born* inflow. So, what's going on?

Well, notice from the graph that the number of rabbits killed per fox falls to a new, lower steady-state value. It's declining because average rabbit speed is increasing. As it does, rabbits being munched by foxes falls—because there's a fixed number of foxes, and their rabbit kill rate is declining, therefore the total volume of rabbits being killed by foxes must fall. And because it falls, the *being born* flow is able to exceed the sum of the two outflows—recall that the *being munched* flow was carrying the lion's share (or was it fox's share?) of the outflow from the rabbit population. Because the inflow to the rabbit stock exceeds the sum of the two outflows, the population begins to grow. And as it does, the volume of both the *being born* inflow and the *naturally dying* outflow do likewise. But, the former goes up by 0.25 times the number of rabbits, while the latter goes up by only 0.05 times the number of rabbits (the birth and death rates of rabbits, respectively). This means that the inflow will be growing a lot faster than the sum of the outflows—one component of which (i.e., *being munched*) is actually declining!

Draw Conclusions & Assess Robustness of Conclusions

We're now at the step in the Model-construction process called Drawing Conclusions. I'll combine that step with the one that follows it: Assess Robustness of Conclusions.

We can conclude that the model, in its most recent incarnation, generates an evolution in the attribute of foot-speed, and that it does so in a manner consistent with the notion of a natural selection pressure. However, the simulation results are not very robust. In particular, as a result of the increase in its average foot-speed, the rabbit population is able to grow *without limit*. In a real ecosystem, this would not occur. A challenge to students in a follow-on assignment, or in an interactive classroom discussion might be to "fix" this new craziness that has arisen.

A good way to motivate the thinking on this score is to ask: *So what would keep the rabbit population from increasing without limit in the real ecosystem*? Some possible answers would be: a limited food supply, a growing fox (or other predator) population, and human intervention. If we invoke the "System as Cause" perspective, we would eliminate, on first-pass, the third option because it is outside the current system boundary. Options one and two would be entertainable because both are limits that are internally-generated by relationships within the current system.

Let's consider option one: a limited food supply. We could add to the model a stock called *Rabbit Food Supply* (consisting of an aggregation of all vegetation in the ecosystem upon which rabbits feed), and then include an outflow—*consumed by rabbits*—and an inflow—*growing*. But, if we were looking for the simplest possible way to include a constraint from rabbit food availability, we would take a simpler approach.

The availability of vegetation would exert an impact on the rabbit *death rate*—it may also influence rabbit *birth rate*, but probably not until things got pretty nasty in terms of rabbit starvation. So, in order to keep things simple, I will include an impact from food availability on only rabbit *death rate*. And, rather than add a stock to the system, and then have to formulate and numerate new flows, I will proxy the effect of food by relating the associated impact to the size of the rabbit population. That is, as the number of rabbits climbs above some *threshold* value, the impact on rabbit *death rate* will grow increasingly severe. If you are sensing the emergence of a graphical function, your Systems Thinking instincts are sharpening!

Figure 9-15 shows the next incarnation of the model, and Figure 9-16 depicts the resulting dynamics.

(see Rabbit6 in the *Intro to Systems Thinking* folder on your disk*)*

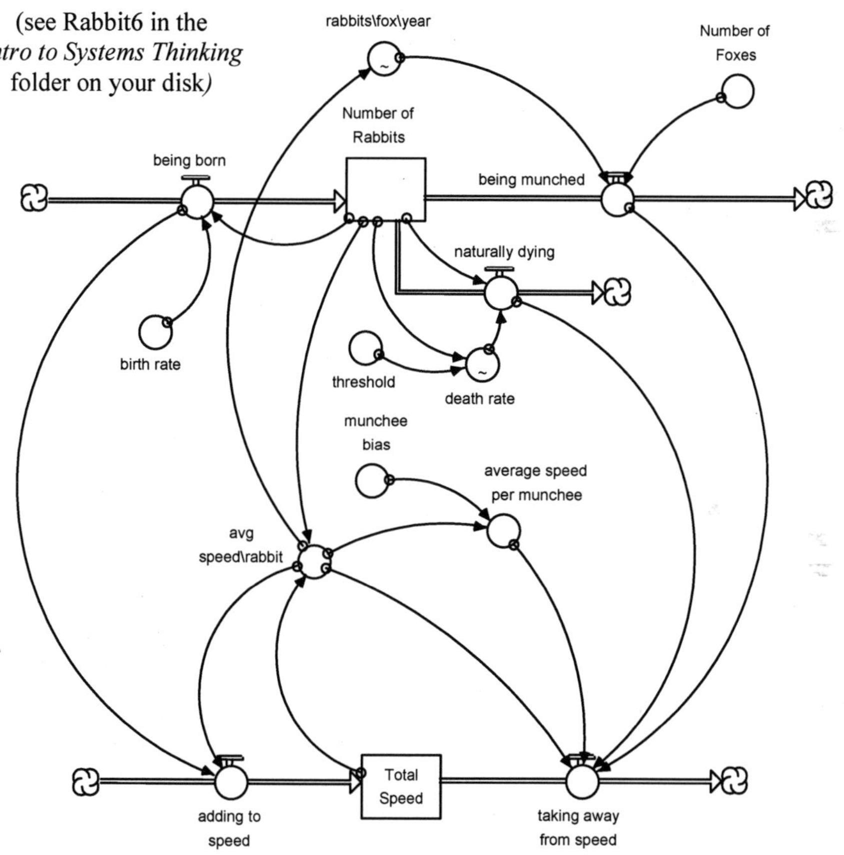

Figure 9-15.
Model Number Six.

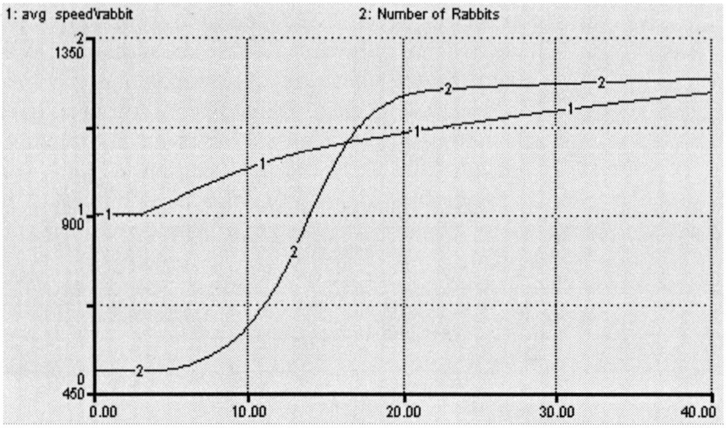

Figure 9-16.
Simulating Model Number Six.

As you can see, when rabbit *death rate* increases as the rabbit population increases beyond some threshold level, the rabbit population no longer grows without limit. Instead, it exhibits the classic S-shaped pattern characteristic of most animal and plant populations growing in ecosystems in which human intervention has not undermined natural, self-regulating controls. However, it appears that in the process of addressing the rabbit population growth issue, we've re-introduced a problem we had fixed once before. The average speed of the rabbit population seems, again, to be growing without bound.

Actually, that's not the case, and though it appears that we've re-introduced the problem we saw before, the cause of the increase in average speed per rabbit is different in this case. Previously, when average rabbit speed was growing without bound, the issue was that we had omitted the connection from the *being munched* flow to the outflow from *Total Speed*. As a result, rabbits were exiting the population and taking with them, in effect, an average speed of *zero*! This will result in the average speed of the rabbit population approaching infinity over time.

In the current incarnation of the model, the average speed of the rabbit population is approaching a value of 2.0 (if you'd like to confirm this, set the ending time for the simulation of the Rabbit6 model to 200 years, or so, and then simulate). What's going on now is that foxes will catch any rabbit whose average speed is less than the value at which foxes, on average, can run. If you examine the *rabbit\fox\year* graphical function, you will discover that at an average rabbit speed of 2.0, foxes will catch *zero* rabbits per year. At this speed, the rabbit population is in effect under no selective pressure from fox predation—because the average rabbit can outpace the average fox. The average speed of the rabbit population will thus approach 2.0. At this value of average rabbit speed, essentially no rabbits will be munched, and the *being born* inflow will be completely offset by the *naturally dying* outflow—where the rabbit *death rate* will have increased to a value of 0.25 (so as to exactly offset the rabbit *birth rate*).

The implication here is that predation obviates the need for starvation. If foxes limit the population, rabbits will have plenty of food. The more foxes are neutralized as a rabbit population regulator (due to increasing rabbit foot speeds), the more starvation must pick up the slack of controlling the rabbit population. "You pays your money, you takes your choice." We'd have to ask rabbits what balance between predation and starvation they'd prefer. Another option would be to convince rabbits that family planning was a good thing. One way or the other, the rabbit population will be brought into balance with the carrying capacity of the ecosystem.

At this point, students could be asked to conduct a thorough sensitivity analysis of the parameters in the model to determine the robustness of the associated behavior modes and conclusions. The model is, in fact pretty robust as it now stands—given the choice of a model boundary (e.g., assuming the fox population is constant and does not co-evolve its foot-speed).

In Conclusion

This Chapter has illustrated the progression of steps in the modeling/learning processes using developing an "extension" exercise as a context. The specific activities you execute would vary as the context changed, but the set of steps in both processes would remain the same.

What's Next

Chapter 10 will conclude the Guide by providing a set of Guidelines for executing each step in the modeling/learning processes using "best practices" that we've discovered over a couple of decades of constructing these kinds of models.

Chapter 10

Guidelines for the "Writing" Process

Both writing and constructing a model, are inherently *creative* processes. Creative work, by definition, is not something you produce by simply adhering to a prescribed set of guidelines associated with a well-defined set of steps. This said, after teaching Systems Thinking, and using it with clients for more than twenty years, I can say with great confidence that there is a set of steps—and a set of guidelines/principles of good practice associated with those steps—that "work." By "work," I mean that if someone follows them, the likelihood they'll produce a model that underwrites understanding, inspires insight, and guides effective action, increases significantly. It makes sense to read the material in this Chapter carefully, and to keep it nearby as a reference when you engage in modeling activity.

The Chapter uses the "steps" in the modeling process (the framework presented in Chapter 8, and illustrated in Chapter 9) to organize the presentation of guidelines and principles of good practice. Here are the "steps," Model-construction first, then Learning Process.

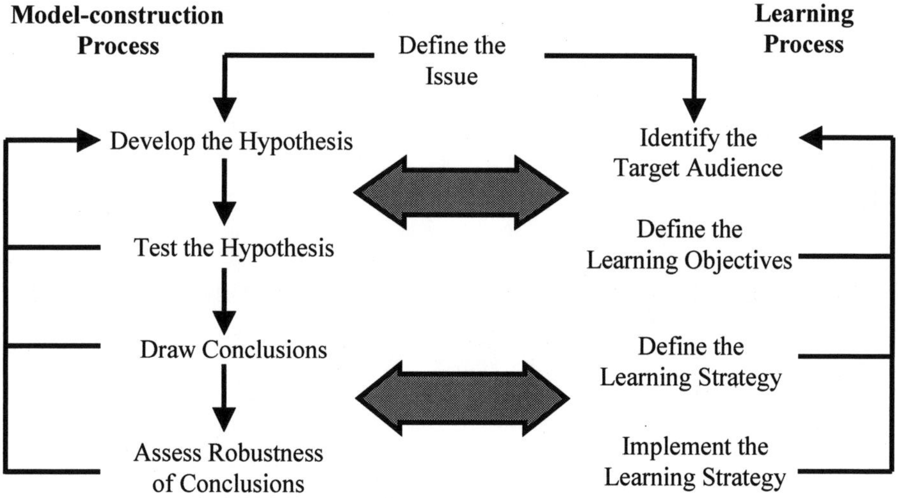

Figure 10-1.
The "Steps" in the Model-construction/Learning Processes.

The Model-construction Process

Define the Issue

Most modeling efforts that spin out of control do so because not enough time was taken up front to nail down a clear statement of purpose. Spend the time!

Guideline 1: Write a Clear, Explicit Purpose for the Effort

- Write your purpose statement on paper. Then, always keep it within sight.
- Couch your purpose statement in terms of gaining an understanding of the relationships responsible for generating a specific dynamic phenomenon. For example: "The model is intended to shed light on the causes of revolutions as a general phenomenon."
- Never set out to, or be drawn into, "modeling the system." Always focus on understanding a phenomenon.

Guideline 2: Develop a Reference Behavior Pattern

A *Reference Behavior Pattern* (RBP) is a graph over time of one or more variables that best depict the pattern of behavior you're trying to understand.

- In cases where there is "history," your RBP should show "As Is." Your RBP may also include a "To Be" segment. In developing the "To Be" segment, pay particular attention to how long it is projected to take to bring about the change.
- Choose an "interesting" RBP, one that visually depicts "a puzzle" that *cries out* for an explanation.
- Where possible, create the RBP by using *relative* measures. Divide the RBP variable by some benchmark quantity, to screen out issues of absolute growth (this is called "normalizing").
- Pay attention to the time span over which the RBP is unfolding. Only include things in your model that unfold with a rhythm that's relevant to this time span. For example, if the RBP unfolds over five years, you could include things that play out in quarters or months, but not weeks or hours!

Two illustrative Reference Behavior Patterns appear in Figure 10-2. The first (a) shows the total number of crimes committed by youths over a certain number of years. The second (b) shows the percentage of crimes committed by youths that were violent. Notice that the second pattern is a lot more thought provoking than the first. The total number of crimes committed by youths could be rising over time simply because the number of youths is rising. But, if a "normalized" variable (like any percentage) is increasing, it isn't due to an *absolute* increase in another variable. In this case, it's due to the fact that the *nature* of crimes being committed by youths is changing. That's interesting! It stimulates creative thought...exactly what a *Reference Behavior Pattern* is designed to do.

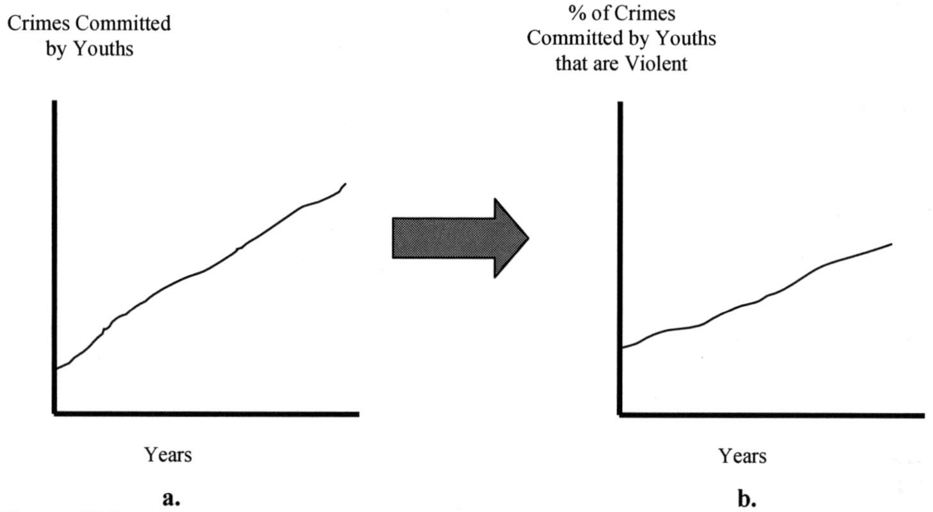

Figure 10-2.
Two Illustrative Reference Behavior Patterns.

Develop the Hypothesis

In this step, you will conceive of, and render, a hypothesis that you believe is capable of explaining the *Reference Behavior Pattern.*

Guideline 1: Seek a Dynamic Organizing Principle

A "dynamic organizing principle" is an infrastructure-based, or feedback loop-based, framework that resides at the core of your model. Think of it as providing an underlying theme for your "story."

Some examples: ***Infrastructure-based***

- Time allocation
- Main Chain (Chapter 7)
- Attribute tracking (Chapter 7)
- Relative Attractiveness (Chapter 7)
- Slippery/Sticky Perceptions (Chapter 7)

Some examples: ***Feedback loop-based***

- Overshoot and collapse (Chapter 7)
- S-shaped growth (shift in dominance, Chapter 6)
- The two-stock pure/dampened oscillator structure (Chapter 6)

Guideline 2: Try: Main Chain, Key Actor, Most-important Accumulation

Main Chain

A Main Chain is a sequence of stocks connected by conserved flows. The concept of a Main Chain infrastructure is described in Chapter 7. A Main Chain provides a physical "backbone" off of which the remainder of your model can pirouette. To arrive at a Main Chain, ask: *What's flowing in this system?* Then, note the "stages" through which it is flowing. If the associated accumulations/flows form a conserved-flow sequence, you've got a Main Chain. Figure 10-3 illustrates a Main Chain.

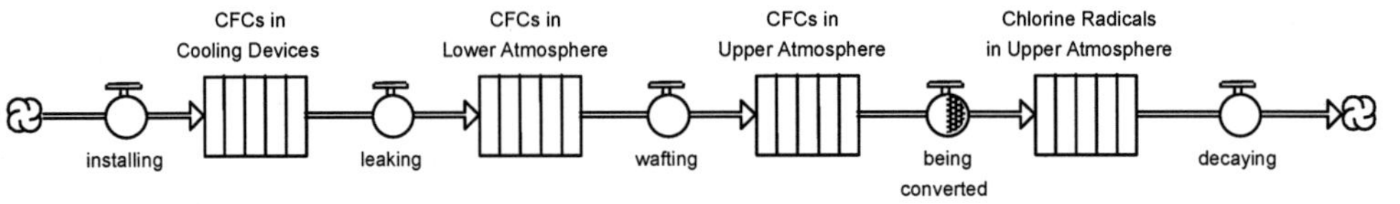

Figure 10-3.
An Illustrative Main Chain.

"Key Actor" Approach

Often, by thinking in terms of the "actors" associated with your issue, you can identify a nucleus of essential stocks and flows associated with each.

- Identify the smallest set of Key Actors that you hypothesize to be involved in generating the RBP. Actors usually are not individual human beings.
- For each actor, identify the *conditions* the actor monitors to determine how things "are" within their piece of the system. Conditions may be material (e.g., money) or non-material (e.g., trust). They will most likely be represented as *stocks*.
- Next, identify the *actions* taken by each actor in response to changes in conditions. The actions will be represented by *flows*.
- Finally, identify the *resources* that support taking the actions. Resources may be material or non-material. Resources most likely will be *stocks*.
- It's useful to employ a "Key Actor Matrix" to collect the information on *conditions*, *actions*, and *resources* (illustrated in Figure 10-4).

Actor: *A Plant Population in an Ecosystem*

Conditions Monitored	Actions Taken	Resources Consumed
Water	Stomata opening/closing	Energy
Nutrients	Draw down inventory of stored nutrients	Energy

Figure 10-4.
An Illustrative Key Actor Matrix.

The Most-important Accumulation

Identify the accumulation (a stock) you consider to be closest to the heart of the issue you are seeking to address. Then, add an inflow and an outflow. Proceed to Guideline 3.

Guideline 3: Use The Generic Flow Templates to Characterize the Flows

Once you've got some stocks and flows laid out, the next step in developing the hypothesis is to *characterize the flows*. Seek to capture the nature of each flow as it works in reality. Strive to achieve an operational specification by using one of the generic flow templates described in the Appendix to Chapter 5.

- Look at each flow in isolation. Think about the nature of the activity the flow is representing. Do not ask: *"What are all the factors that influence this flow?"* That question leads to a Laundry List!
- Ask: *"Is the flow stock- or flow-generated?"*
- If it's flow-generated, use the Co-flow template.
- If it's stock-generated, ask: *"Is it the stock to which the flow is attached that's doing the generating?"* If the answer is yes, use either the Compounding or Draining template, depending on whether it's an inflow or outflow. If an external stock is doing the generating, use the External Resource template.
- If it's a perception process, use the Stock-adjustment template.
- Use only one template per flow. If more than one template is needed, create a separate flow for each. Add only the structure

that's part of the generic flow template you choose. You can embellish the structure later.

Guideline 4: Close Loops Without Adding More Elements to the Model

After you have characterized the flows in your model, the next step in rendering your hypothesis is to close loops—but without including additional stocks, flows, and converters to your model.

- Look to see if any of the parameters associated with the generic flow templates should depend on some other variable currently in the model. Many should, but you're only interested in representing the interdependency if doing so is part of your hypothesis! Close the relevant feedback loops, as illustrated in Figure 10-5.

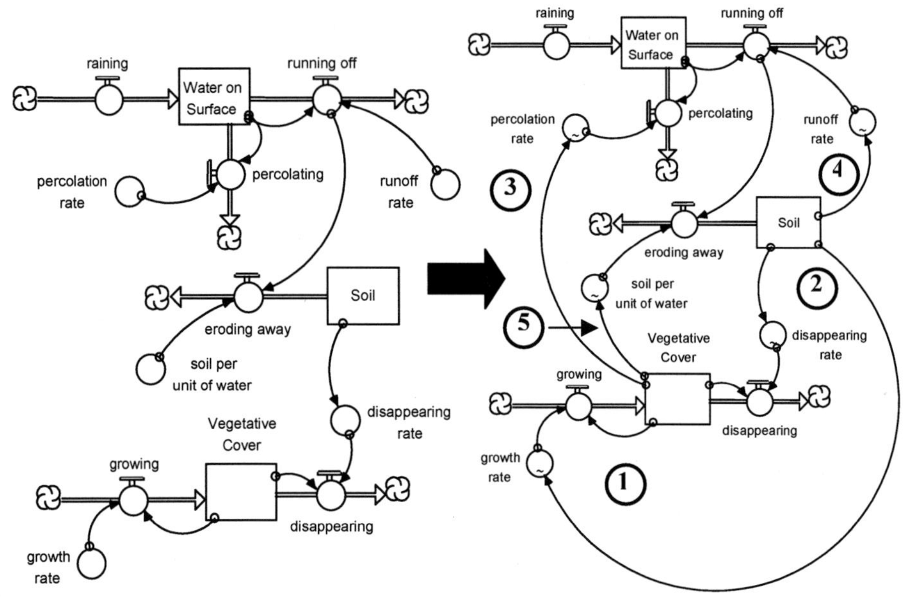

a. Open-loop Generic Flow Templates **b. Closing Loops Within the Templates**

Figure 10-5.
From Open to Closed-loop Generic Flow Templates.

In Figure 10-5a, five generic flow templates are used to specify the flows in a system designed to look at erosion issues. Three draining, one co-flow, and one compounding template are used. Each of the five parameters associated with these templates is a constant. In Figure 10-5b, loops are closed by running wires from the relevant stocks to each of the formerly-constant parameters (each link that closes a loop is numbered in the Figure).

Guideline 5: Click in Equation Logic

Once you have fleshed out the map, select a chunk (a sector is a good chunk, if you are using sector frames) and make it simulatable. Begin by clicking-in the associated algebra...

- Click-in the generic flow template equations first.
- Check the dimensional balance of each equation. The units-of-measure of the right-hand side of your equation should be the same as those for the left-hand side. Flows should have the same units-of-measure as the stocks to which they are attached, but with the addition of "per time."
- Avoid "dead buffalos" (a name derived from their appearance on the diagram; see Figure 10-6). Strings of factors that are "correlated," or "influence," are not the same thing as an operational statement of causality. Avoid "Critical Success Factors" Thinking.

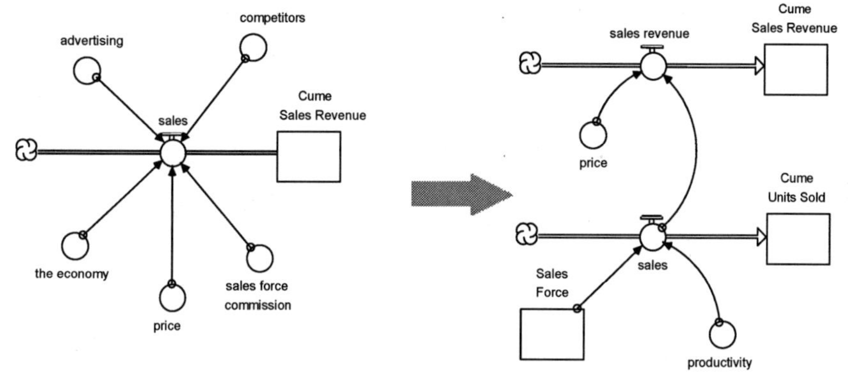

a. "Dead Buffalo" **b. An Operational Specification**

Figure 10-6.
A "Dead Buffalo" vs. an Operational Specification.

Guideline 6: Numerate

On first-pass, choose numerical values that initialize your model in steady-state. In steady-state, the net of all inflows and outflows across each stock is zero. See the Appendix to this Chapter for details on, and an illustration of, initializing a model in steady-state.

- On first-pass, choose numbers that are simple and make sense relative to each other. Simple, internally-consistent numbers will help you to put a model into steady-state. On second-pass, you can include real-world data, if that is important to your purpose.

- Favor small numbers (e.g., 0, 1, 10, 100, 1000, etc.) over large ones (e.g., 2.7182818, 97.2222, 1 quadrillion, etc.). By choosing small, numbers, you'll find it much easier to understand what's going on in your model.

- Avoid equations in which more than two to three "effect" or "impact of..." multipliers are strung together. Even if each multiplier has a value of only slightly less than 1, the resulting overall impact of several strung together can be surprisingly large (e.g., 0.9 * 0.9 * 0.9 * 0.9 = 0.66).
- Follow the procedure outlined in the Appendix to Chapter 6 for developing your graphical functions.

Test the Hypothesis

Simulating your hypothesis on a computer is designed to increase your confidence that the model you've rendered is useful for the purposes it is intended to serve. These tests also are designed to make you aware of your model's limitations. Thus, you should emerge from this step both confident in what your model can do, and aware of what it can't.

In order to ensure that you learn as much as you can from each test, before each simulation, sketch out your best guess at an outcome, and be explicit about the rationale for the guess. Then, after the simulation is complete, work to resolve any discrepancies in either actual versus predicted behavior, or in your rationale. If test results so dictate, don't hesitate to cycle back through Steps 1 and 2 of the modeling process.

Guideline 1: *Find & Fix "Mechanical" Bugs*

- Simulate. Then, investigate and eliminate any "?" (i.e., undefined entities) that prevent your model from simulating.
- Once the model simulates, choose Range Specs from the Run Menu. Look for any anomalous values (?, ∞, and negative values that should be positive values).
- Put any "offending variables" and associated inputs (or inflows and outflows) into a Table. Set the Table's print interval to DT. Set the Table to report Beginning Balances.
- Simulate for a few DTs. Run your eye over the values in the Table to determine which variable(s) is causing the problem.
- Repeat the previous two steps until you have identified and eliminated all mechanical mistakes.

Guideline 2: *Ensure that Model is in Steady-state*

- Enter the stocks onto a graph. Simulate. You should see all straight lines (except for some variables that may have extremely small scales and exhibit microscopic fluctuations).
- Put any stock(s) not in steady-state, as well as their inflows and outflows, into a Table or Graph. Trace through the problem (you can use the "T" on the bottom of the Graph page to trace through the logic using hover pop-up graphs). Repeat, as needed, until the model is in steady-state.

Guideline 3: *Test "Robustness" of the Model*

- Robustness tests identify formulations that do not hold up "under extreme conditions." They also reveal inherent dynamic tendencies.
- Incorporate STEP and PULSE functions as test-inputs into one or two flow equations. The idea is to "shock" the associated stock, knocking it out of its steady-state resting place.
- Graph the response of key variables. In particular, graph the response of the stock whose flow is delivering the shock.
- Use the tracing feature (the "T" on the bottom of the Graph page that appears after selecting a variable name in the Graph header), in conjunction with the "hover to display mini-graph" feature, to help you understand why you are getting the results you're getting.
- Determine whether the model is exhibiting absurd or implausible behavior: stocks going negative when they shouldn't; stocks growing without limit; system returning to the "wrong" steady-state; response time too short, too long, etc. *Be sure you understand why you are getting the results you are getting.* Often you will have omitted a feedback loop, or one is there, but is too weak or strong.
- If the system exhibits a "high frequency" oscillation in response to any test (i.e., stocks and/or flows jump up and down wildly each DT), check your DT. Halve its value, per the guidelines presented in the *DT: What, Why & Wherefore* section of the *Help Files* within the software. If the oscillation persists, halve it a few more times.
- If the system exhibits a smooth, but ever-expanding, oscillation pattern, be sure that you're using one of the Runge-Kutta simulation methods. See *Simulation Algorithms* in the *Help Files* within the software for details.

Guideline 4: *Replicate RBP*

- Check model results against the *Reference Behavior Pattern.*
- First-pass, look for qualitative similarity. Be sure you understand, and can explain, the results you are generating.
- Use the tracing feature (the "T" on the bottom of the Graph page that appears after selecting a variable name in the Graph header), in conjunction with the "hover to display mini-graph" feature, to help you understand why you are getting the results you're getting.
- If required to "track history," substitute in real-world numbers for constants, initial values and graphical functions. If necessary, also add time series graphical functions to drive the model with historical data.

Draw Conclusions and Assess Robustness

The ultimate purpose of constructing and simulating a model is to draw some conclusions. Once you have done so, it is important to assess their robustness—that is, under what "conditions" do the conclusions hold (there are precious few universal truths!)?

"Conditions" can be broken into two categories: *behavioral assumptions* (which reside in parameters *internal* to the model boundary) and *scenario assumptions* (parameters *external* to the model boundary that determine the nature of the environment outside the system).

A conclusion's robustness depends on how well it "holds up" under a range of variation of behavioral and scenario assumptions, and under different choices with respect to model boundaries. There is no "official standard" of robustness. If your conclusions remain the same under a reasonable range of variation of both kinds of parameters, you can consider them robust. *You* have to define "reasonable."

What's perhaps more important than "certifying robustness" is for you to be aware of under what conditions your conclusions cease holding up. And, when they do, what *new* conclusions emerge? It's also important for you to understand why conclusions stop holding up, and why new ones emerge when they do. Having a firm grip on the answers to these kinds of questions ensures that you have a mature appreciation for the conditions under which what you've concluded applies, and does not apply.

Guideline 1: Conduct a Thorough Sensitivity Analysis

- Begin with Behavioral assumptions.
- Break out, as a constant, a "base/normal" value from any graphical function in which this value is "buried" within the function.
- Subject all constants, one at a time, to a rigorous sensitivity analysis using the *STELLA* software's Sensitivity Analysis functionality.
- Next, subject sets of constants to same treatment.
- Vary initial conditions of key stocks using Sensitivity Analysis—again, first one at a time, then sets.
- Move to Scenario assumptions. Repeat the preceding.

Guideline 2: Challenge the Model Boundaries

- Challenge the Extensive and Intensive Model Boundaries (see Figure 10-7).
- Examine each cloud associated with a material flow. Ask: *What would happen if I covered up the cloud with a stock?* If the mental simulation suggests an interesting possibility, add the stock, the associated flows, any resulting feedback linkages, numerate, then simulate. See if your conclusions hold up.
- Look at the most interesting stocks in your model. Consider each as a candidate for disaggregation. Run some mental simulations. If the results appear to suggest something interesting, pursue it.
- Select any constants and/or graphical functions in the model that were shown, through Sensitivity Analysis, to be "sensitive" (i.e., capable of causing conclusions to change). Disaggregate them

(i.e., use stocks, flows and converters to represent them at a more detailed level). See if they lose their "sensitiveness."

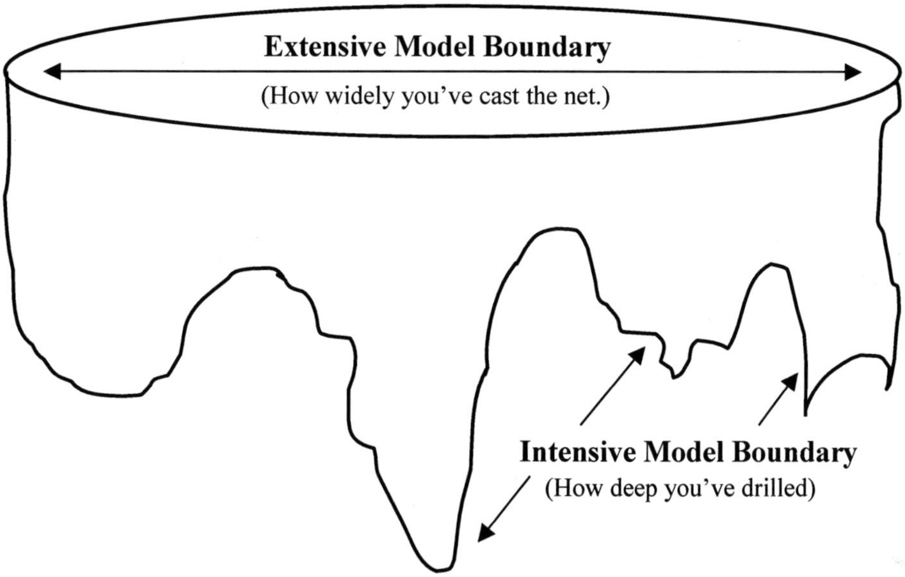

Figure 10-7.
The Concept of an Extensive and Intensive Model Boundary.

The Learning Process

Define a Target Audience & an Associated Learning Strategy

If only the person/people who construct a model learn from it, a huge potential for increasing understanding and insight will go un-harvested. It is therefore vitally important to define an overall learning strategy for your modeling effort. The universe of strategies breaks into two types: *active* and *passive*. The *STELLA* software is designed to support the former. This means using the software to make what you have learned "available" for other people to learn through their own "discovery-oriented" learning process. The Guidelines provided here assume this approach.

If you teach, at whatever level, one key target audience is students.

Guideline 1: Use the Continuum to Structure Student Learning Experiences

Recall the continuum of *STELLA*-based exercises presented in Chapter 8 (Figure 8-5). At the far left is Exercising. At the far right is Constructing. And, in the middle, is Extending. In employing *STELLA*-based exercises with your students, proceed from left to right: *Exercising* to *Constructing*. This will enable students to gain familiarity with language, concepts, and good modeling practice before having to do any conceptualization or model-boundary setting of their own.

- *Exercising* "products" include: *STELLA* models with and without interfaces, Flight Simulators (with built-in coaching), Virtual Laboratories, and Electronic Storybooks.

Guideline 2: Follow Guidelines for Developing Coaching Sequences

Coaching Sequence Guidelines

- Establish learning objectives.
- Create a decision-making "game" or performance challenge that has a well-defined target objective(s). Target audience members should find the game fun and enjoyable, yet challenging. The game should enable learners to capture the learning targeted in the learning objectives.
- Determine which learning objectives will require coaching sequences in order to be realized. Provide NO coaching on the first simulation. In general, front-end load "in-character messages," then shift away from messages and toward coaching sequences as the simulation run number increases.
- Determine the prevalent behavior patterns generated by the model in response to learner decisions. The patterns fall into two classes: those that constitute "winning" (good performance), and those that constitute "losing" (crash and burn).
- Make certain that each such pattern is "covered" with some sort of message and/or coaching sequence. Make sure "winning" is covered by an acknowledgement of "good performance."
- All coaching sequences should be delivered "JITJWN" (Just In Time, Just What's Needed). No lectures! No long "time outs."
- Coaching sequences should be designed to "catch learners with their mental models up." One of the best ways to do this is to employ IF-THEN-ELSE logic to "sense" when they have made a particular decision. Decisions, at least those involving thought, are all generated by mental models. This means you can infer what the learner "must have been thinking, in order to have arrived at that particular decision, under those particular conditions."

  Surface for the learner "what they must have been thinking" at the outset of the coaching sequence. Simple word-and-arrow diagrams, especially if animated using Flash™, are a great way to do this.

- Then, using the "here's what you must have been thinking" diagram as a departure point, modify it to illustrate an "enhanced" way of thinking about what's going on. If the original thinking is a straight line of cause-and-effect, close a loop, show an unintended consequence, include a significant delay, or add a variable not included in the straight-line sequence.
- Suggest to learners how the enhanced mental model logic can help them to make better decisions in the game—which will lead to better performance.
- Provide a "final debrief" after the learner has "won."

Guideline 3: Use Storytelling with Chunk-by-chunk Simulation

If your audience is something other than students (though the following works for students, as well)...

- Never present the full-blown structure of a model. Instead, use Storytelling (refer to the *Online Help Files*) to unfurl it chunk-by chunk.
- Annotate your "stories" from your "readers'" viewpoint, *not* from your own!
- Mix simulations in with the unfurling so that learners build their understanding of structure, then behavior; a bit more structure, then modified behavior...and so forth.

Summary

The major benefit of working systematically through the modeling/learning process is that, as a result, you will be able to communicate more clearly, more succinctly, and with greater confidence, both about what's causing the issue you've examined and what you might do to address it in an effective manner. You'll also be able to make your learning available to others in an *active* format. They'll be able to discover insights and build understanding for themselves.

What's Next

Getting out there and doin' it! The time has come to take what you've learned and use it in the world to construct better mental models, simulate them more reliably, and communicate them more effectively. We'll be looking for your "short stories" on the NY Times Best Seller list!

Appendix:
Initializing Models in Steady-state

Achieving a steady-state initialization often is a straightforward process. It's always a useful process to engage in, for two primary reasons. First, steady-state ensures that the parameters in your model are internally consistent. Second, once the model is in steady-state, you'll be able conduct controlled experiments in which you observe the "pure" response of the system to your robustness and policy tests.

In steady-state, the sum of the inflows for each stock is equal to the sum of the outflows for each stock. Therefore, the magnitude of all stocks will be constant.

Guidelines

- Determine how much latitude you have for determining the value of each flow. Can you *directly* set a value for the flow, or otherwise cause it to take on whatever value you'd like?
- Once you've determined the amount of freedom available to you, use the data which are "most solid" to infer values for parameters whose values are "least solid." Give yourself plenty of license in determining "less solid" parameter values. Because few real systems are in steady-state, you'll often need to modify numbers taken from an actual system in order to achieve a steady-state. After completing steady-state-based tests, you can substitute "real" values back in. Whenever possible, use algebraic initialization to establish initial values for stocks. The model in Figure 10-10 will be used to illustrate this process.

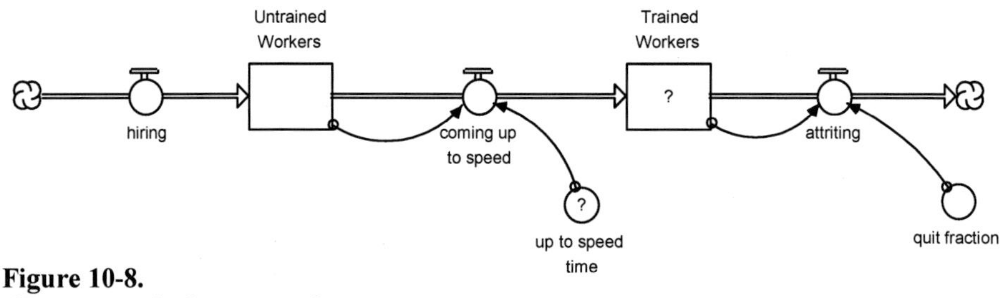

Figure 10-8.
Illustrating Algebraic Initialization.

You may find it useful to work along with this illustration. The model is contained in your *Models* folder as "Initialization." In this simple model, we know the following with some confidence:

- *Untrained Workers* = 100

- *coming up to speed* = *Untrained workers / up to speed time*
- *attriting* = *Trained Workers* * *quit fraction*
- *quit fraction* = 0.25
- *hiring* = 50

We'll use what we know with confidence, to solve for the two values we're less certain of (an initial value for *Trained Workers* and a value for *up to speed time*). The solution process will also cause the model to be initialized in a steady-state condition. We'll begin by finding an initial value for *Trained Workers*. In steady-state...

- *attriting* = *coming up to speed*

Substituting the algebra for the two flows, we have...

- *Trained Workers* * *quit fraction* = *Untrained Workers / up to speed time*

Solving for *Trained Workers*, we get...

- *Trained Workers* = *(Untrained Workers / up to speed time)/quit fraction*

[We can now click-in this equation as an initial value for *Trained Workers*, using variables in the Allowable Inputs list of that stock's dialog box. The equation will then be solved by the software, once, at the outset of the simulation, to arrive at an initial value for *Trained Workers*. Note that, because of the way we solved for this value, it will cause the *attriting* flow to equal the *coming up to speed* flow!]

To wrap up the process, we'll find a steady-state value for *up to speed time*. For steady-state:

- *hiring* = *coming up to speed*

Substituting the algebra...

- *hiring* = *Untrained Workers / up to speed time*

Substituting in numbers...

- 50 = 100 / *up to speed time*

Solving for *up to speed time*...

- *up to speed time* = 100/50 = 2

Following this sort of process is a nice way to use numerical values about which you are reasonably confident to "force out" values for which you have little or no information. At the same time, it yields a steady-state initialization for your model. Two birds with one stone!

List of Figures

Chapter 4

Chapter 5

Chapter 6

Chapter 7

Chapter 8

Chapter 9

Chapter 10

Index

D

E

F

G

T

U

V

W